CAMOUFLAGE OF FIXED INSTALLATIONS

U. S. ARMY

Fredonia Books
Amsterdam, The Netherlands

Camouflage of Fixed Installations

by
U.S. Army

ISBN: 1-4101-0914-3

Copyright © 2006 by Fredonia Books

Reprinted from the 1958 edition

Fredonia Books
Amsterdam, The Netherlands
http://www.fredoniabooks.com

All rights reserved, including the right to reproduce this book, or portions thereof, in any form.

CAMOUFLAGE OF FIXED INSTALLATIONS

		Paragraph	Page
PART ONE	LARGE SCALE CAMOUFLAGE PLANNING		
CHAPTER 1.	INTRODUCTION		
	Purpose	1	3
	Scope	2	3
	General	3	3
	Camouflage Detection	4	4
	Camouflage Planning	5	5
	Geographic Location	6	5
	Types of Fixed Installations	7	5
2.	CAMOUFLAGE PLANNING		
	Estimate of the Situation	8	12
	Essential Reference Data and Aids	9	12
	Analysis of the Area	10	13
	Formulating the Preliminary Plan	11	13
3.	CAMOUFLAGE OPERATIONS		
Section I.	Preparation		
	Schedule of Operations	12	15
	Experimental Areas	13	15
II.	Development		
	Layout Grid Control	14	16
	Marking the Area	15	17
	Aerial Checking	16	17
III.	Control		
	Ground Control	17	17
	Clearing	18	18
	Debris and Spoil	19	20
	Traffic Control	20	20
	Equipment and Supplies	21	20
IV.	Inspections		
	Progressive Inspections	22	20
	Ground Inspections	23	20
	Aerial Inspections	24	20
	Night Inspections	25	20
	Materials	26	21
PART TWO	CAMOUFLAGE METHODS AND CONSTRUCTION		
CHAPTER 4.	METHODS		
	Techniques of Patterning and Screening	27	22

		Paragraph	Page
	Two-Dimensional Patterning	28	22
	Three-Dimensional Patterning	29	25
	Screening	30	45
5.	REAR AREA INSTALLATIONS		
	Camouflage of Buildings	31	50
	Individual Buildings and Tents	32	50
	Grouped Buildings	33	58
	Land Communications	34	66
	Concealment of Water	35	81
	Deliberate Minefields	36	82
6.	MAINTENANCE		
	General	37	84
	Specific Recommendations	38	84
INDEX		....	86

PART ONE
LARGE-SCALE CAMOUFLAGE PLANNING
CHAPTER 1
INTRODUCTION

1. Purpose

This manual provides a guide for personnel responsible for planning and camouflaging fixed installations.

2. Scope

This manual covers methods of camouflaging fixed installations such as airfields, supply facilities, transportation facilities, buildings, and other objects and installations of a permanent or semipermanent nature. The material presented herein is applicable to nuclear and nonnuclear warfare.

3. General

The basic principles of camouflage, factors of recognition, construction methods, and the geographic factors effecting camouflage outlined in FM 5-20 are applicable to the camouflage of fixed installations, the primary difference being larger and more detailed camouflage construction of a more permanent nature. Terms used in this manual are general. For example, the camouflage of buildings is described in general terms to include all types of buildings regardless of their use or purpose, because the use or function of the installation being camouflaged has only an indirect bearing on the type of camouflage except where a specific problem is presented. When planning camouflage for a specific building the activity occupying the building will, in many instances, dictate changes in the application of camouflage. In the same way the surrounding terrain will have a direct bearing on the type of camouflage used. The camouflage of areas, concentration of activities, and emplacements for active defense are also described in general terms. Consequently it will be necessary for the camouflage planner to assemble the necessary information from this text to guide him in the designing of camouflage for a specific combination of buildings, areas, or emplacements.

4. Camouflage Detection

Camouflage detection consists mainly in discovering the items that camouflage is intended to conceal. Deliberate camouflage can be subjected to deliberate study of aerial photographs, infrared detectors, and by radar.

a. Aerial Photography. Camouflage detection can be *successfully* accomplished through aerial photograph interpretation. An expert photointerpreter (PI), given time for a careful study of aerial photogrgaphy can produce remarkably accurate results. As an example, an expert PI can recognize objects as small as baseballs on photographs taken from 40,000 to 50,000 feet under favorable conditions with an automatic camera that has a long focal-length lens. Photo interpretation can provide accurate data on the shape and size of objects.

   (1) *Shape.* The shape of surfaces and objects can be determined exactly by the use of stereoscopes and the study of shadows. A great deal of information can be deducted from typical shapes (tanks and elevators) or typical groupings of installations or elements of an installation, such as tank farms, depots, rail, and industrial complexes.

   (2) *Size.* The size of an object can be measured to within inches of actual size in good vertical stereopairs; and can be estimated within 10 percent of actual size on oblique photos. For additional information on photo interpretation, see TM 30-245 and TM 30-246.

b. Infrared Detection. There are two general types of infrared dectors. Active (near) infrared which requires illumination of the target by some light source i.e., infrared spot or flood lights or the sun. The other type is a passive (far) infrared detector which detects infrared reflectance emissions of the target and converts the signal to a visual picture graph or sound signal. The latter passive device can observe in complete security since it emits no detectable signal.

   (1) Active (near infrared concealment depends on the reduction of contrast between the target and its surroundings or background. Thus, if target and the surroundings are of the same infrared reflectance and texture total concealment is achieved.

   (2) Passive (far) infrared concealment is dependent on reducing heat emission of objects which are hotter than their surroundings. Therefore some insulator or shield must be employed. Defilade, heavy brush or tree cover will attentuate the heat radiation dependent on its density or thickness.

c. Radar. Radar concealment depends on the reduction of radar reflection of installations. Generally speaking, reflection of an installation may be reduced through the elimination of its angular characteristics or through the use of defilade or by digging-in. Foliage cover is ineffective against radar detection.

5. Camouflage Planning

The camouflage of any large area or installation requires careful planning. Plans are to include provisions for seasonal changes, terrain pattern changes, and present and possible future uses of the installation.

6. Geographic Location

A fixed installation, the geographic position of which is public knowledge, cannot be camouflaged against air-to-ground or guided missile attack because the bombing run, or the homing factor of the missile, will be the geographic position itself. Therefore, any attempt to camouflage the object or installation will be of no avail.

7. Types of Fixed Installations

For the purpose of this manual a fixed installation is considered to be any installation which may remain in one location long enough to enable the enemy to observe, locate, and attempt to destroy it. In some situations a specific installation will be considered to be fixed; and in other situations when compared to other types of installations, it will be considered to be mobile. For instance, the command post (CP) of a battalion is fixed in relation to the companies of that battalion, and the same CP is mobile in relation to army HQ. This varying degree of stability calls for varying degrees of camouflage effort. For the purposes of this text, the following installations will be considered as fixed installations.

a. Established Geographic Position. An installation which is a permanent part of the culture or industry of a locality, such as the Pentagon 1, (fig. 1), and Washington National airport 2, (fig. 1), and whose geographic position is familiar to everyone is called an established geographic position. Paragraph 6 gives the main reasons why this type of position cannot feasibly be camouflaged.

b. New Geographic Position. An installation built as a result of a wartime emergency and which is so far to the rear as to be out of the combat zone, such as a temporary Army camp in the U. S. during World War II, is, for the purpose of this manual, called a *new geographic position*. Since this emergency site can be planned in advance it is possible to conceal it successfully with camouflage construction, but only if the camouflage effort precedes the establishment of the installation. In other words, it is normally too late to think of applying camouflage once the first changes are made in the appearance of the terrain. But if large-scale camouflage construction precedes the actual construction at the site it will be possible to conceal the location from hostile observation.

c. Large TOE Installation. A supply or service installation in the army service area or communications zone and of semipermanent nature falls in this category and can be concealed by camouflage. Large TOE installations include supply points, communications centers, and transportation terminals or transfer points. The camouflage methods and techniques to be used are determined basically by two factors:

    (1) Whether the camouflage is planned and accomplished prior to the establishment of the site:

    (2) Whether camouflage must follow the actual occupation of the site (fig. 2), due to the tactical situation or other demands 1. Figure 2 illustrates an airfield before camouflage and 2, figure 2 illustrates the same airfield after completion of camouflage. The agriculture pattern has been maintained during construction and camouflage of the airfield. The farm building complex painting on the camouflaged runway is included for deception.

d. Routes of Communications. Routes of communication, such as roads, trails, railroads, bridges, communications centers, transportation facilities, and ferry points require varying degrees of camouflage depending on the tactical situation. A road which is known by the enemy to exist and which is being used as a main supply route (fig. 3) may be screened to conceal the amount and type of traffic using it. The roads serving an Army airfield (fig. 4) will require complete concealment when the very existence of the airfield is being concealed.

e. The Command Post. The battalion, regimental, or divisional command post may be a semipermanent installation and should be camouflaged as well as the situation permits. The camouflage of a command post shown in figure 5 is a good example of camouflage for this type of installation.

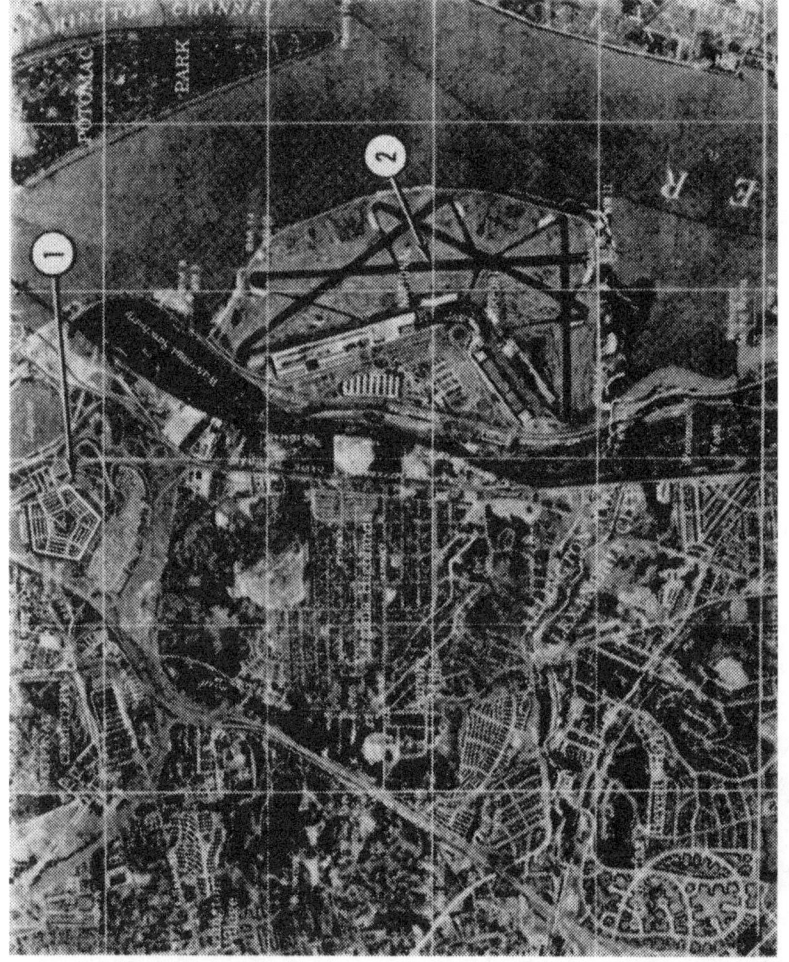

Figure 1. Established geographic positions.

1 Pentagon. 2 Washington National airport.

1 Before camouflage.
2 After camouflage.

Figure 2. Example of camouflage of deliberate army airfield

Figure 3. A main supply route screened by a 1,650-ft net

Figure 4. Airfield access roads requiring camouflage

Figure 5. Camouflage matching type of terrain (mottled).

CHAPTER 2
CAMOUFLAGE PLANNING

8. Estimate of the Situation

In making an estimate of the situation leading to the formulation of a camouflage plan, the following factors must be considered:

a. The Objective of the Project. The mission of the installation, equipment, or activity will affect the plan as to the degree of camouflage effort to be expended, and the objective of the camouflage itself, whether it is to be designed to hide from observation, to deceive the observer as to the real purpose, or to be a combination of both of these.

b. The Priority of the Project. The urgency of the situation will indicate a local priority, although higher authority may prescribe other priorities based on overriding factors. After being assigned a time-schedule for completion, estimates will be made to determine the time required for the various phases of the project and a project time-schedule will be established.

c. The Area and Location. The factors of geographic position are to include a consideration of the extent of the project in square miles; the analysis of landmarks or reference points, and the determination of whether the location is unique or is one of several similar type installations.

d. Coordination. It should be determined whether coordination will be required with other projects in the same general area, and whether the project will become a part of an overall deception plan. In addition coordination with civilian defense agencies should be considered.

e. Cooperation. The necessity for cooperation with other units, units of other services, other allied armed forces, local residents, or public utilities personnel must be determined.

f. Effort to be Expended. Determination should be made as to whether the camouflage project is designed to be accomplished by troops alone or whether the assistance of contract labor is available. It is also necessary to know how much money can be expended and whether the camouflage is of a temporary nature, or is to be maintained over a period of years.

g. Availability of Materials. It should be ascertained if the camouflage will be limited to what can be accomplished by indigenous or natural materials. The availability of flat-tops of cotton twine garnished nets and garnished wire netting must be determined if artificial materials are to be used. If artificial materials such as lumber and building materials are to be used, necessary priorities for their procurement are to be furnished.

9. Essential Reference Data and Aids

The materials which will be accumulated prior to the establishment of the plan, and which will be used as reference and aids in the determination of the plan are:

a. Topographic Maps. Medium- and large-scale topographic maps of the immediate and adjacent areas.

b. Aeronautical Charts.

c. Photomaps and Mosaics. A controlled mosaic, at a scale of not less than 1:25,000 (fig. 6), of the project area; and a controlled or semicontrolled mosaic at a scale of not less than 1:50,000 of the surrounding area should be provided. For additional information on photomaps and mosaics see FM 21–26.

d. Aerial Photographs. Aerial vertical photographs of the project area with a minimum overlap of 60 percent, and oblique photographs, taken from cardinal directions or most likely approach angles.

e. Town Plans and County Maps.

10. Analysis of the Area

a. Landmark and Reference Points. A study is to be made to determine the existing landmarks and reference points from which an enemy aircraft can locate a camouflaged area.

b. Possible Decoy Sites. A study of the adjacent area is to be made to determine the existence of possible sites for decoy camouflage installations designed to deceive enemy planes, infrared detectors, and radar.

11. Formulating the Preliminary Plan

Conclusions drawn from the assembled data and the study of the area will lead to a preliminary camouflage plan. A more detailed plan will be developed after the preliminary plan is approved.

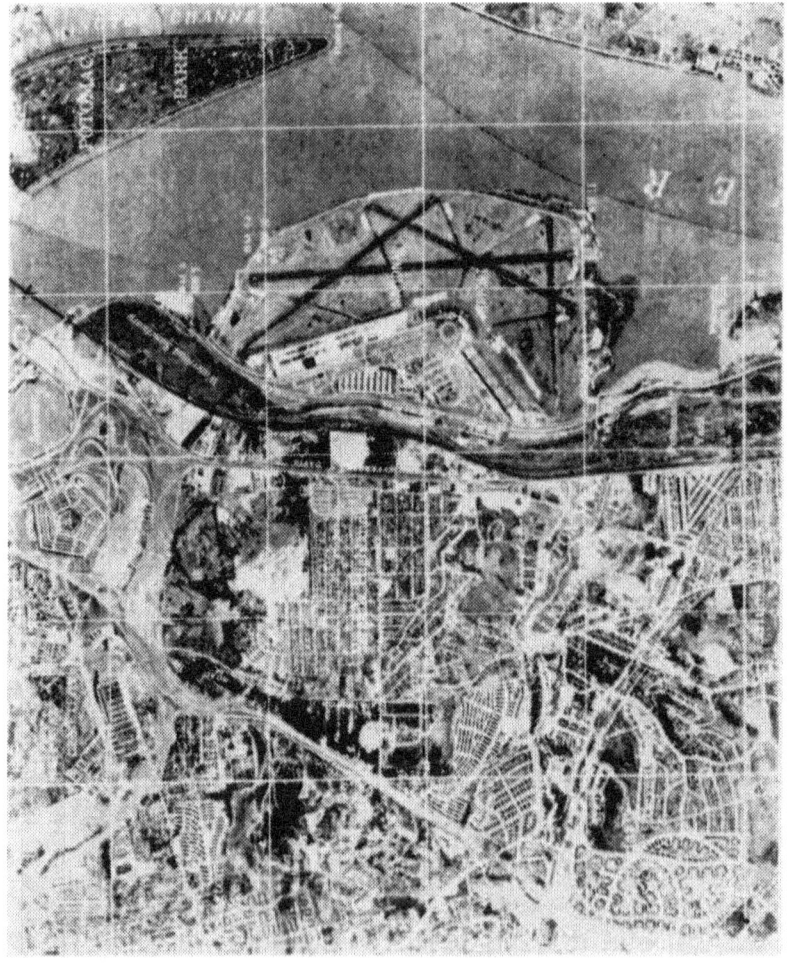

Figure 6. Large-Scale (1:25,000) mosaic of area shown in figure 1

CHAPTER 3
CAMOUFLAGE OPERATIONS

Section I. PREPARATION

12. Schedule of Operations

A schedule of operations is to be prepared early in a camouflage project.

a. Planning the Schedule. The schedule should be planned so that:
   (1) The project does not interfere at any time with the functioning of the installation.
   (2) Materials can be ordered and deliveries scheduled to avoid a storage problem.
   (3) Different types of work do not conflict by being carried on at the same time in the same area.
   (4) One type of the work can be substituted for another in case of unforeseen delays. As an example it may be possible to do construction work such as landscaping and carpentry, simultaneously, and at the same time accomplish limited 2-dimensional and 3-dimensional patterning. However the situation may prevail where the landscaping must be completed prior to other activities.

b. Factors Affecting the Schedule. The factors which affect schedule completion are:
   (1) Type and availability of labor and equipment.
   (2) Delivery of materials.
   (3) Weather.
   (4) Extent of work required.
   (5) Combat conditions.

13. Experimental Areas

It is frequently desirable in rear area installations to set up experimental areas to test colors and techniques designed for proposed camouflage methods for large installations. The final appearance of a camouflage plan cannot always be visualized from research data or design drawings. The only proof that the plan is the correct one is observation and study of full-scale work. This must be done periodically from the start of the project to prevent waste of time and effort on unsuitable plans and to permit necessary changes in materials, design, color, or construction.

Section II. DEVELOPMENT

14. Layout Grid Control

a. The Modular System. A plan of the area to be camouflaged should be drawn at a scale of not less than 16 feet to the inch, and should be gridded at 4-foot intervals using the modular system. The 4-foot interval, or module, is a unit of measurement for regulating proportions. Its use as the basis for a plan will reduce building costs, offer the designer a simplified method of dimensioning drawings, eliminate the necessity for much expensive detailing, and offer a system of repetitive modules 4 feet in dimension on which repetitive aspects of the camouflage can be constructed by an assembly-line type of production. The 4-foot interval coordinates the sizing of different materials on a common basis so that when assembled they can be readily fitted together to form a complete structure. The better the different components from different manufacturers can be fitted together, the less will be the cutting and adjustments required on the job. Planning for the use of modular products does not hamper designers in creating camouflage construction to meet any need. It simply means that designers, producers of building products, builders, and craftsmen all work together on a common basis using a coordinated system of dimensioning.

b. Modular Coordination in Drafting. Preparation of working drawings on a modular basis is not essentially different from that customarily followed in architectural practice. However, a new factor has been added—the discipline of the grid.

(1) *The modular grid.* Coordination of building products in a structure is based upon a 4-inch cube represented as a 4-inch grid on plans, elevations, and sectional drawings (1, fig. 7).

(2) *Small-scale drawings.* At scales of less than 1 inch to the foot it is not practical to show grid lines. An architect's scale permits drawings to be laid out in multiples of 4 inches (2, fig. 7). Plans and elevations for camouflage construction are to be laid out using a 4-inch grid.

(3) *Modular details.* The 4-inch grid is used in drawing typical details at a scale of 3 inches or 1½ inches equals 1 foot. The grid is the basis of coordination and not necessarily a dimension of materials. Materials are

shown as actual size and either located on, or related to, a grid line by a reference dimension. Dimensions on grid lines are shown by arrows; those not on grid lines by dots (2, fig. 7).

d. Reasons for use of 4-inch Module. A 4-inch module was chosen because:

(1) It is large enough for manufacturers to reduce the number of stock sizes and still satisfy consumer demand.

(2) It is small enough for ample freedom in design and for flexibility in equipment layout.

(3) It coincides with the dimensions of a great many building materials already standardized and is applicable to present construction practices.

(4) It is a unit of measurement with which architects, builders, masons, and carpenters are already familiar.

(5) It approximates in centimeters (.39 inch) the basis of proposals by countries using the metric system.

15. Marking the Area

After the layout plan is determined a grid of 16-foot squares (4 modules) is transferred to the ground. Lines marked on the ground must not vary more than 1 foot from the design to maintain the scale of the design. Variations in scale make it difficult to match patterns at side walls and roofs and between areas. Lines can be marked on the ground using a tennis court marker, chalk lines, or other marking device. With the gridlines as guides, the pattern outlines are then drawn.

16. Aerial Checking

Once the construction starts, it should be constantly checked from the air. Frequent observation flights should be made and aerial photographs of progress made at regular intervals. If this is done, breaches in camouflage discipline and weaknesses in the camouflage plan can be discovered and corrected.

Section III. CONTROL

17. Ground Control

Camouflage discipline is vital during all construction phases at any site to be camouflaged. Effective camouflage discipline requires constant supervision when construction is designed for a large area. Careless and widespread earth scarring (fig. 8) is caused by failure to establish a track plan for vehicles, by too free

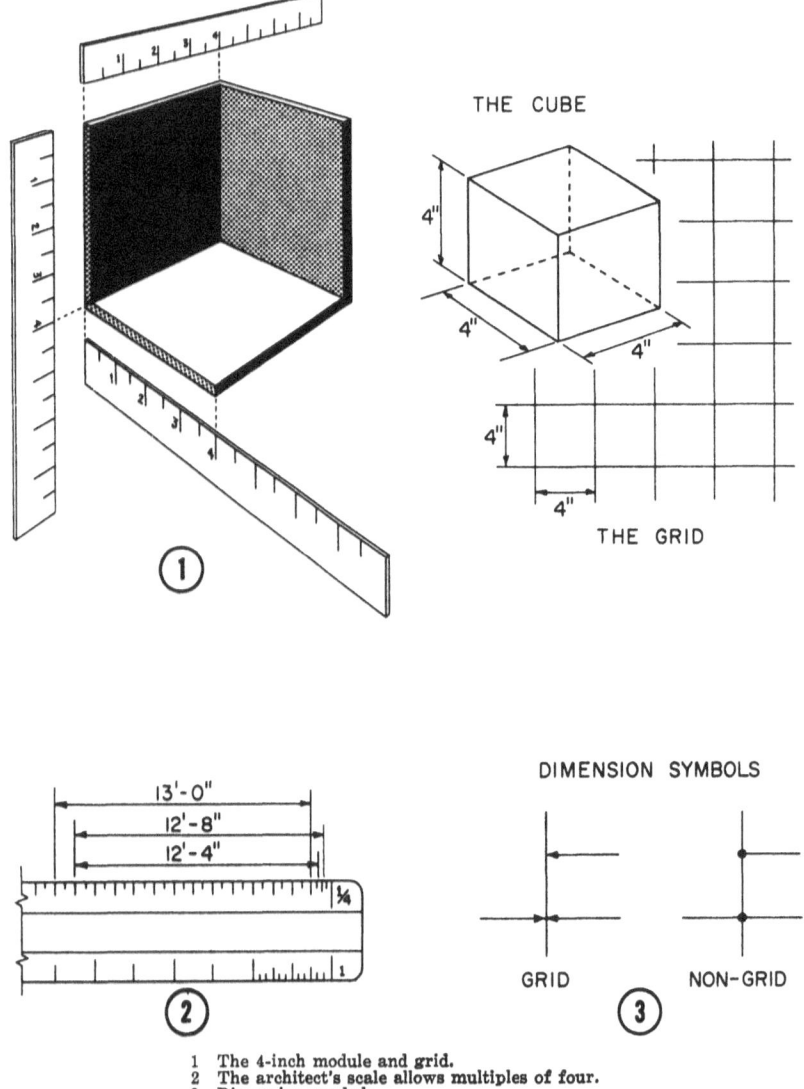

1. The 4-inch module and grid.
2. The architect's scale allows multiples of four.
3. Dimension symbols.

Figure 7. The 4-inch module as used in drafting.

use of earthmoving equipment, and by too extensive leveling and clearing in order to make construction easier.

18. Clearing

When a large area of woods must be cleared, as for an airstrip, trees should be cut along clearings in an irregular pattern rather than in a straight line, even though this requires more than the minimum of clearing (fig. 9). Clearings for buildings, however,

Figure 8. Uncontrolled movement of heavy equipment resulting in a glaring indication of activity

should be limited to the area to be occupied by the building. Building locations should be shifted slightly to avoid cutting down trees that will make subsequent camouflage easier.

19. Debris and Spoil

Scrap lumber, packing boxes, empty paint cans, and other forms of refuse and debris, as well as the spoil from excavations, should be disposed of or camouflaged as soon as possible.

20. Traffic Control

The traffic plan, set up during the design phase, must be enforced with guards where necessary. Personnel and vehicles must stay on assigned paths and roads which must be wired or clearly marked. All personnel must be familiar with the plan as it concerns their activities. Parking areas must be dispersed and concealed, and waiting points and turnarounds must be marked.

21. Equipment and Supplies

Equipment not in use and stockpiles of supplies must be concealed or removed from the site. Working equipment must be screened for security. To prevent scarring the earth around small concrete structures such as pillboxes, raised platforms can be used for concrete mixing, supplies, and spoil.

Section IV. INSPECTIONS

22. Progressive Inspections

During construction, materials must be constantly checked for suitability, quality, color, and proper application.

23. Ground Inspections

The overall construction plan must be checked frequently on the ground for compliance with the design. The application of paints and other materials must be checked frequently. Compliance with camouflage discipline must be enforced by frequent inspections.

24. Aerial Inspections

A check of the work is made by direct observation and by aerial photographs for indirect analysis. Target approach techniques are used to check for landmarks and reference points.

25. Night Inspections

Both direct and indirect observations are made at night from ground and air to discover any violations of the principles of good camouflage.

Figure 9. Irregular clearing for airstrip to expedite camouflage.

26. Materials

The paints used should be checked for color and type. Cotton nets and wire netting must be inspected before use to assure that they are properly garnished as to color and pattern.

PART TWO
CAMOUFLAGE METHODS AND CONSTRUCTION

CHAPTER 4
METHODS

27. Techniques of Patterning and Screening

There are three basic methods of camouflaging large installations; they can be used single or in combination. They are 2-dimensional patterning, 3-dimensional patterning, and screening.

a. Two-Dimensional Patterning. Two-dimensional patterning is applied to flat surfaces like roofs, walls, or the ground. It is frequently used where traffic or observation must not be restricted or obscured. Objects with a depth or height of a few feet can be simulated more convincingly than objects with greater elevations, such as houses. The lower the level of aerial observation, the less effective two-dimensional patterning becomes.

b. Three-Dimensional Patterning. Three-dimensional patterning is applied to all types of surfaces. It is more realistic because it casts shadows and is therefore preferable to two-dimensional patterning.

c. Screening. Screening hides an installation from observation and is effective against both high and low-level observation. It is used when 2- or 3-dimensional patterning does not give adequate concealment. Smoke screening, a special technique, is discussed in paragraph 30*d*.

28. Two-Dimensional Patterning

This method has been used successfully in camouflage construction requiring flat or nearly flat patterns.

a. Disruptive or Pattern Painting. Disruptive or pattern painting is used to "break up" and change the characteristic appearance of a flat surface or a building. 1, figure 10 illustrates an antiaircraft tower before camouflage; 2, figure 10 illustrates the tower after disruptive painting; and 3, figure 10 illustrates an aerial view of the completed camouflage with excellent concealment resulting from blending with the shape of the rocky ledge.

1 Antiaircraft tower before camouflage.
2 Disruptive painting has been applied.
3 Aerial view of tower (circle) showing excellent concealment.

Figure 10. Concealing tower from aerial observation.

(1) Patterns simulating three-dimensional objects can be painted on flat surfaces. In all pattern painting, the controlling factor is appearance from the air.

(*a*) Objects like orchard trees which are tall and close

together almost fuse when seen from an oblique aerial view. Therefore, there should be little spacing between "pattern trees".

(b) Allowance must always be made for shadow. At certain time of day, for instance, the shadow of a hedgerow is about as wide as the hedge is high. Therefore, hedge shadows should be painted the height of nearby hedges.

(2) Patterns having slight depth or elevation are more effective representations of three-dimensional objects than flat patterns alone.

b. Scarification. Scarification of the open land around a large installation gives the appearance of cultivated fields. Breaking up adjacent areas with different types of equipment and running furrows in different directions gives patterns of contrasting textures and tones.

c. Grass Seeding. Grass seeding provides a comparatively quick, natural-colored cover for earth scars around new construction. Also, when it is planted as part of a ground pattern of simulated fields, it gives a variety of color and texture.

d. Controlled Irrigation. When water is abundant and distribution equipment is available, patterned areas may be produced by varying the amount of water used in each area.

e. Controlled Fertilizing. Controlled fertilizing produces the same results as controlled irrigation.

f. Controlled Killing Agents. Common toxic chemicals can be applied to grass to produce patterned areas.

g. Controlled Mowing. Cutting areas of grass, hay, or similar growth to different heights gives varying textures to a field. Varying direction of cutting produces marked changes both in texture and color. One swath, about 6 feet wide, cut across a field of relatively high grass (30 inches) effectively simulates a deep ditch, even to low-flying observers (fig. 11).

h. Spreading Materials. Mulching is often used to tone down scarred earth around buildings. Cast shadows can be obscured by spreading black cinders or other black waste materials beyond the area of critical shadow. These materials may also be used to simulate groves of trees (fig. 12) and roads. Painted cut brush can be piled to look like orchards and hedgerows. 1, figure 12 shows that this installation is both new and military. 2, In the irregular woods patterns are duplicated in the camouflage by spreading texturing materials in patches and rolling sections of

Figure 11. Simulated ditch made by cutting swath and covering with waste oil.

the textured surface as in 3. Buildings are painted to correspond with the mottled terrain pattern.

i. Controlled Burning. By burning grass within set limits, runways or orchards can be simulated for the brief periods before natural growth re-covers the burned area.

j. Texturing. Texturing to reduce the shine from smooth surfaces is ordinarly limited to paved surfaces and to roofs and sides of structures (fig. 13). Different types of texturing granules can be used on different parts of the same surface to produce variations in texture. For additional information on two-dimensional patterning see FM 5-22.

29. Three-Dimensional Patterning

Two-dimensional patterning is usually satisfactory for a large portion of most camouflage projects; it may be necessary to supplement it with 3-dimensional patterning such as false structures, trees, or other terrain features or through the use of screens.

a. False Structures. False structures (fig. 14) for screens or decoys can be constructed so as to be mobile and vari-shaped. Building material, such as adobe, plaster, cloth, garnished wire netting, and heavily painted window screening (fig. 15) can be used to fabricate these structures. When unusual native materials are to be used, time and needless experimentation can be saved by obtaining fabricating information from local inhabitants (fig. 16). Three-dimensional construction is used to make existing installations, buildings, or fortifications resemble entirely different types of structures. Farmhouses built over pillboxes and debris erected over tank cars to resemble a rubbish heap 1, and 2 (fig. 17) are examples of three-dimensional construction.

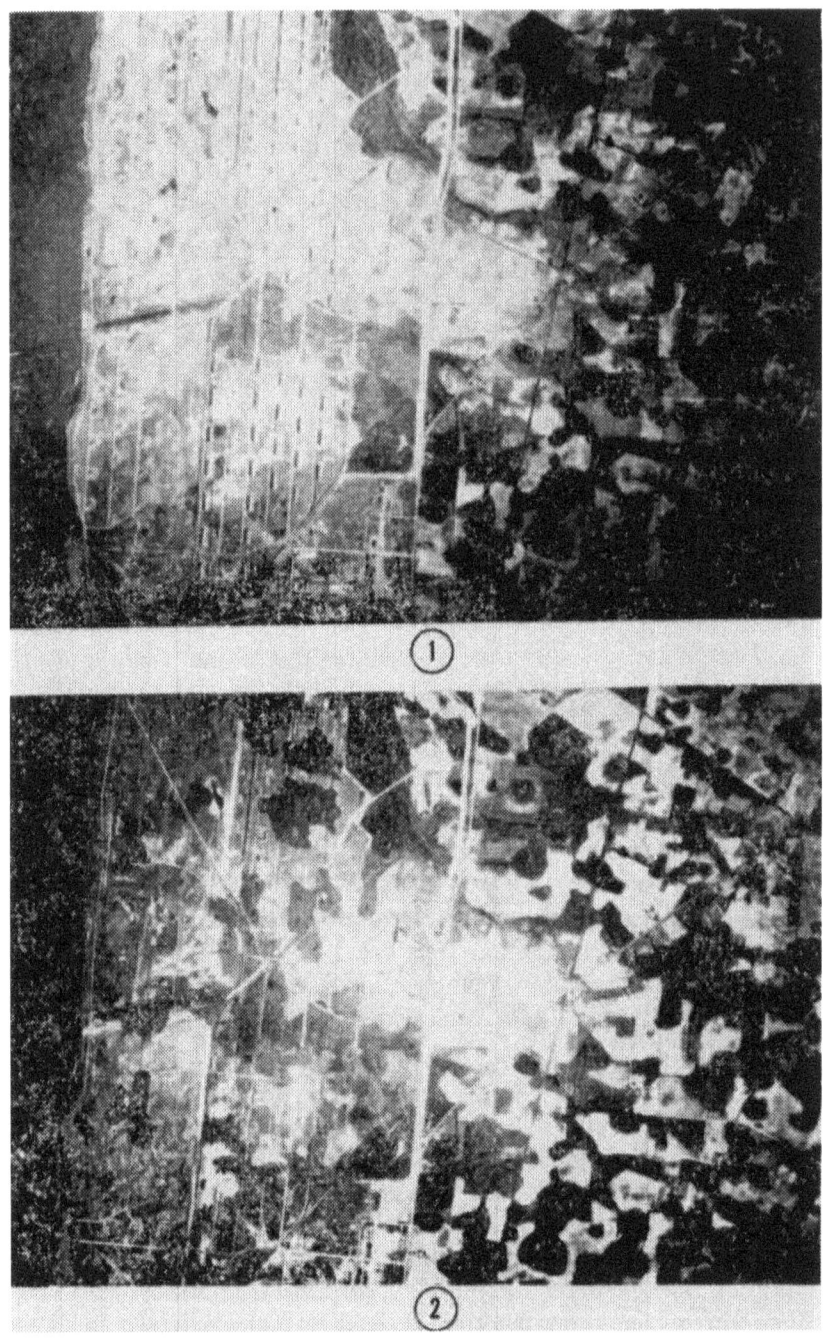

1 Aerial view before camouflage.
2 Aerial view after camouflage.

Figure 12. Camouflage by spreading materials

b. Real and False Trees. Trees can be transplanted in clumps to create dark masses or on the shadow side of new buildings to break up the shadows. False trees can be used to supplement existing overhead cover, to break up regular shadows, or to help disguise cleared areas such as landing strips or runways. A false tree is usually made up of a rigid skeleton on which material is arranged to imitate foliage 1 and 2 (fig. 18). Many variations of false trees and hedges are possible; the most practical types are illustrated in figures 19 through 35. Almost any light material which can be wired or tied to the skeleton to provide texture and shadow will serve as imitation-foliage material. Garnishing strips of osnaburg and burlap, garnished twine nets, salvaged rags, steel wool, feather-garnished wire netting, tin cans (either round or flattened and spray painted), bunches of reeds strung on wire, and palm fronds are all useful. Cut foliage should be used only if it can be replaced continually before it withers.

c. Shrubs. The regular outline of excavations on unscreened emplacements can be broken up by planting shrubs in an irregular pattern along their edges. Shrubs are also effective in breaking up shadows cast by buildings and, if grouped together, can be used to simulate trees.

d. Vines. Quick-growing vines serve several purposes. They may be planted in strips to simulate hedgerows. They may be trained to grow over and hide concrete emplacements and to climb sloping wires or uprights. When set in cans suspended beneath netting, they cover the netting and become on overhead screen.

e. False Terrain Features. Materials may be arranged on the ground to imitate terrain features. Crumpled paper, sagebrush, tumbleweed, or other bush materials may be wired to short stakes in the ground and arranged to simulate truck gardens. Cut hay or straw may be mounded to simulate a dike or revetment. Appro-

3 Spreading and rolling.

Figure 12—Continued.

Figure 13. Texturing materials being applied with an improvised blower

Figure 14. Mobile decoy house.

Figure 15. Full-scale decoy house.

Figure 16. Storage shed constructed to resemble native dwelling.

1 Wood and paper boxes wired to framework.

Figure 17. Simulated rubbish heap for tank car concealment.

2 Inside view.

Figure 17—Continued.

1 Imitation tree with foliage of fibrous vegetation placed in wire netting.

Figure 18. Imitation trees.

2 Foliage of feather-garnished netting, cut and shaped irregularly over framework of wood

Figure 18—Continued.

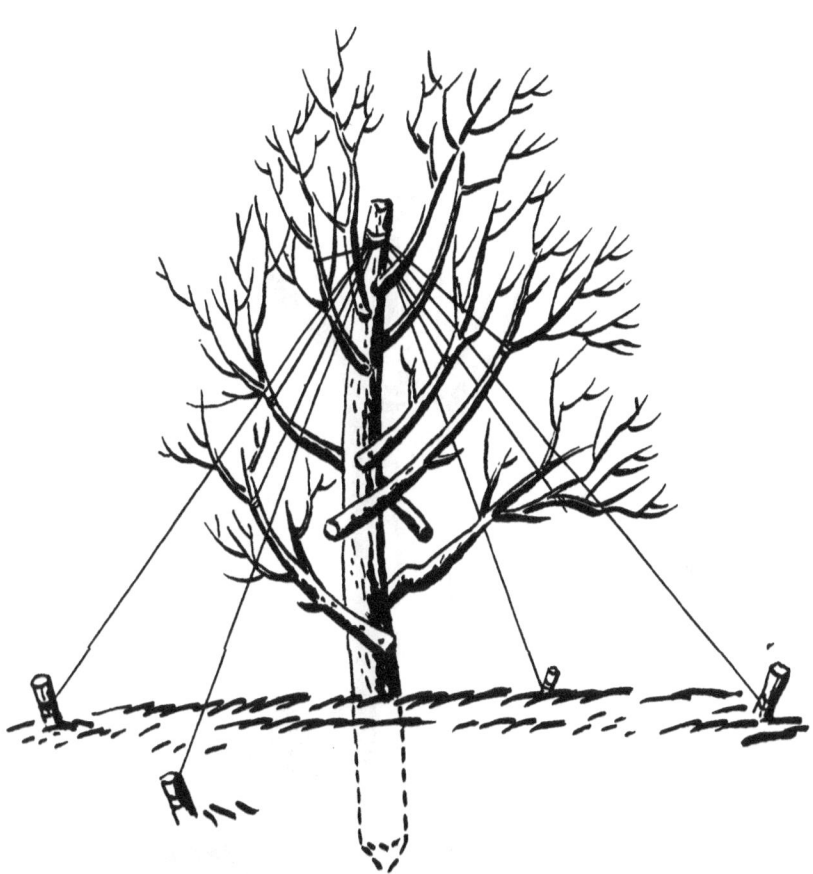

Figure 19. Center post of false tree sunk and anchored.

Figure 20. Method of anchoring center post.

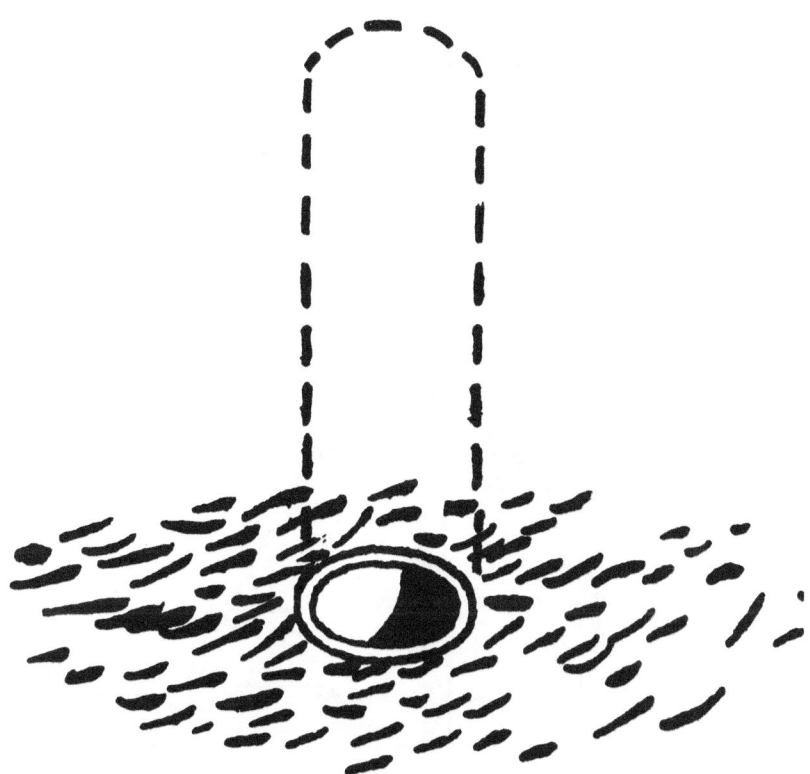

Figure 21. Empty shell canister or pipe used to support false trees.

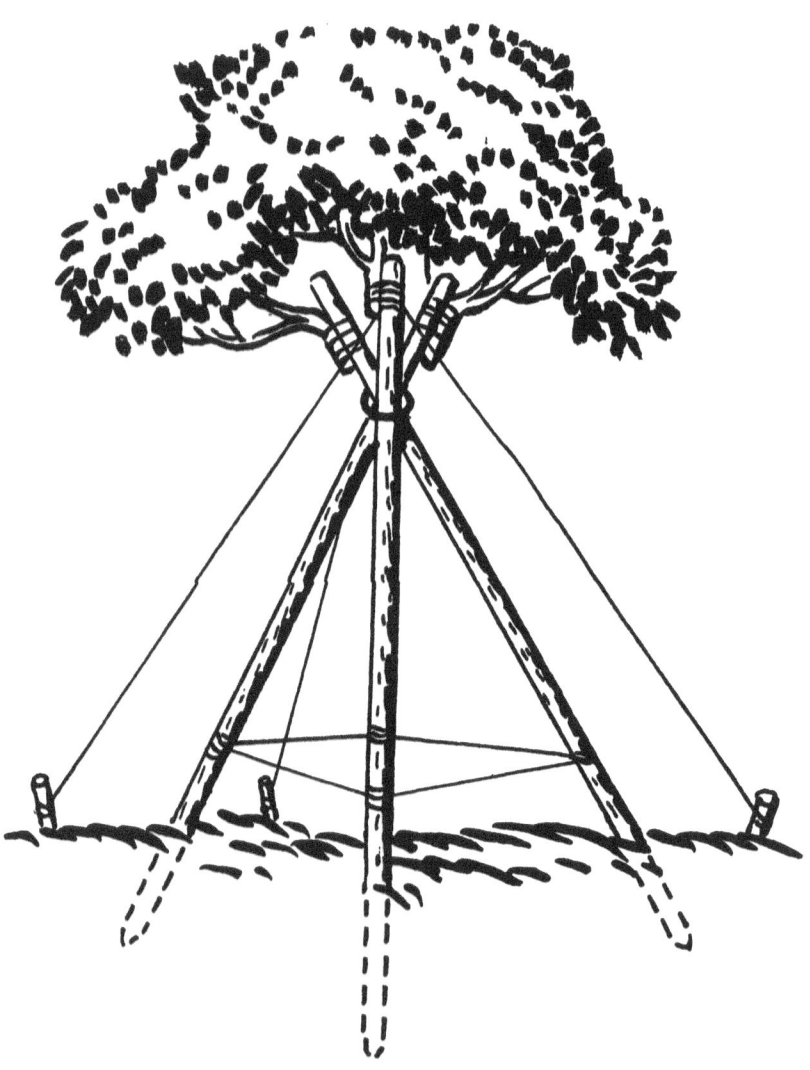

Figure 22. Tripod support for foliage materials.

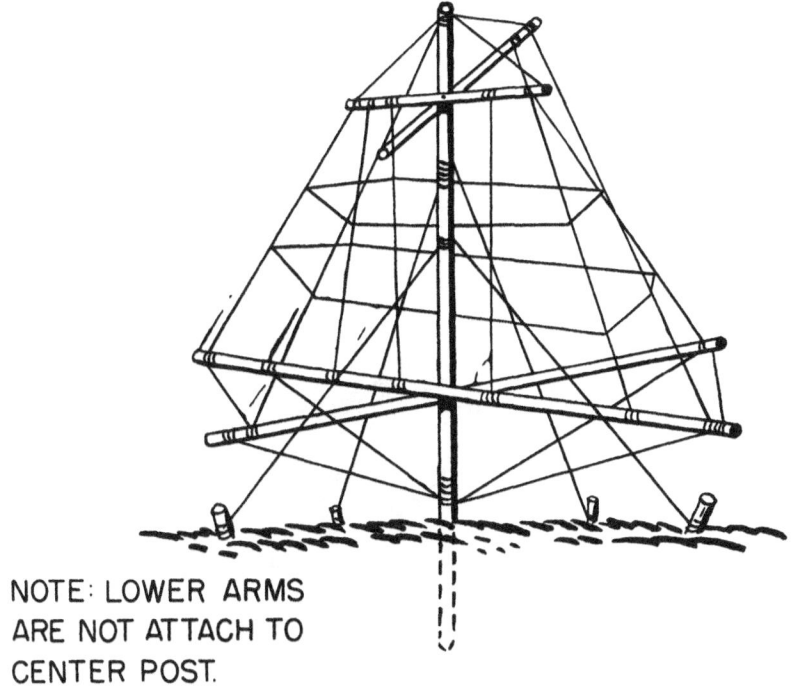

NOTE: LOWER ARMS ARE NOT ATTACH TO CENTER POST.

Figure 23. Framework permitting garnishing to move realistically.

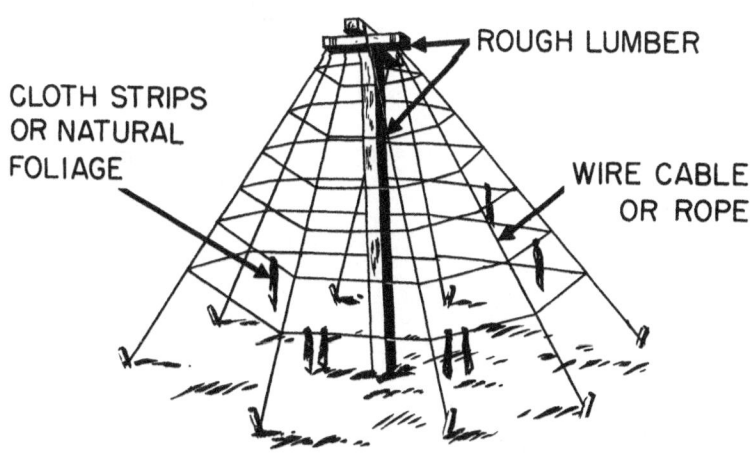

Figure 24. Suitable framework for low, wide tree.

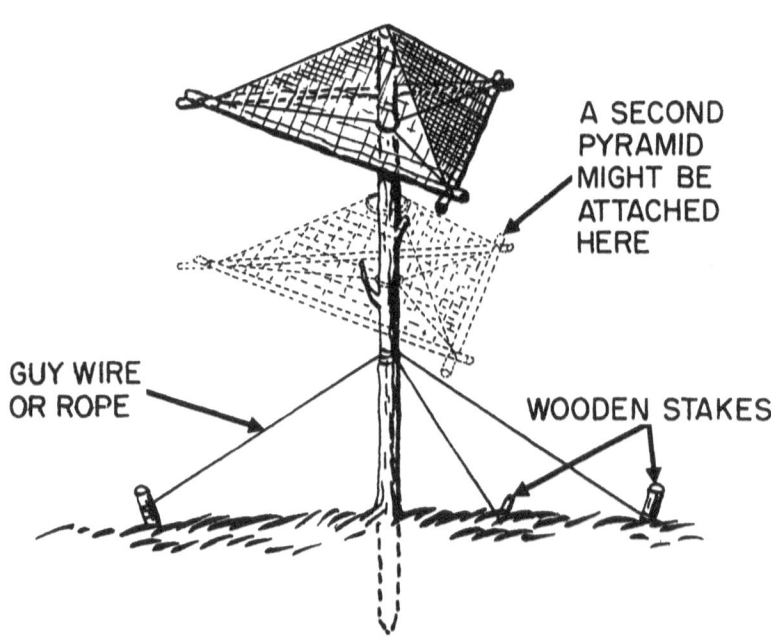

Figure 25. Center post supporting a pyramid of wire netting.

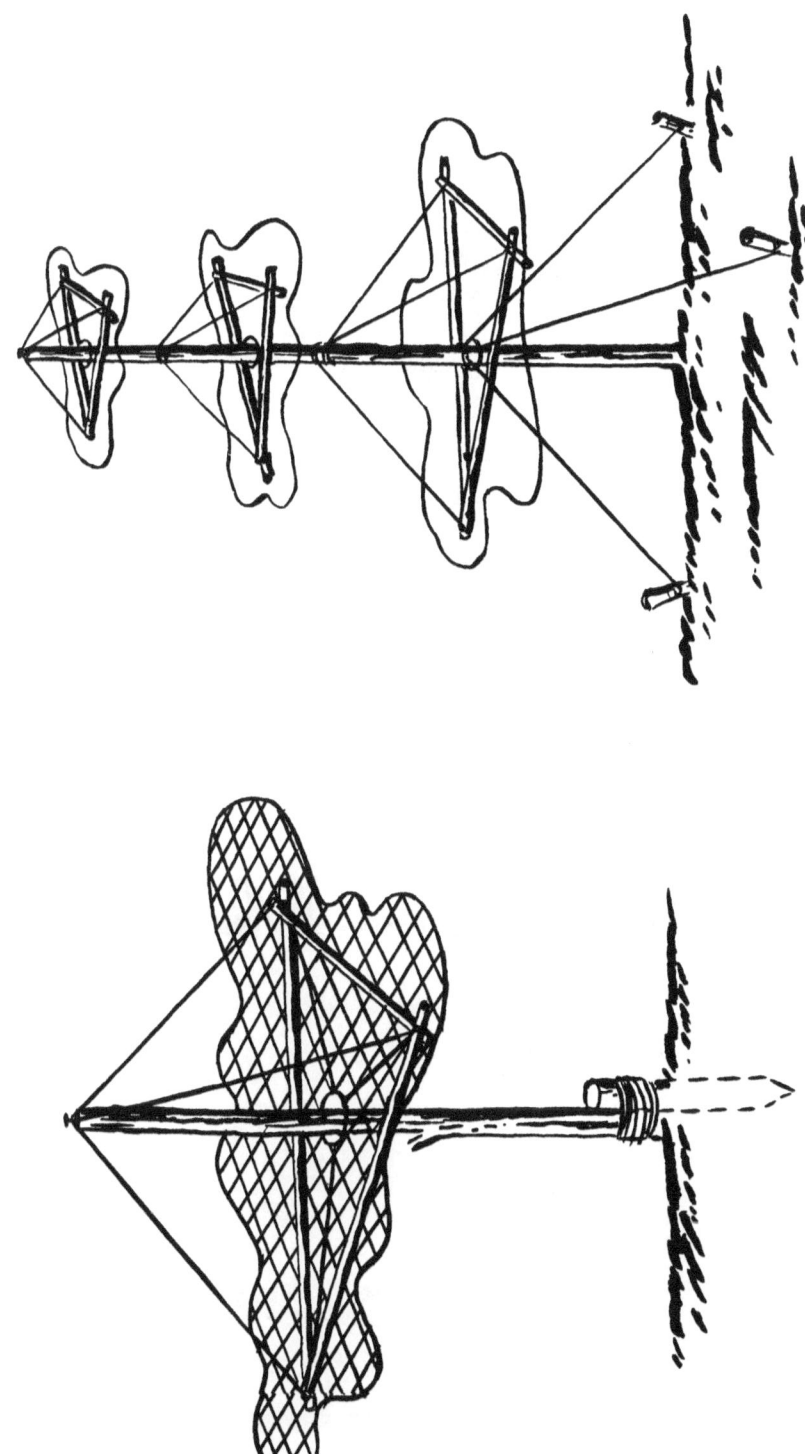

Figure 26. Use of flat panels in false-tree construction.

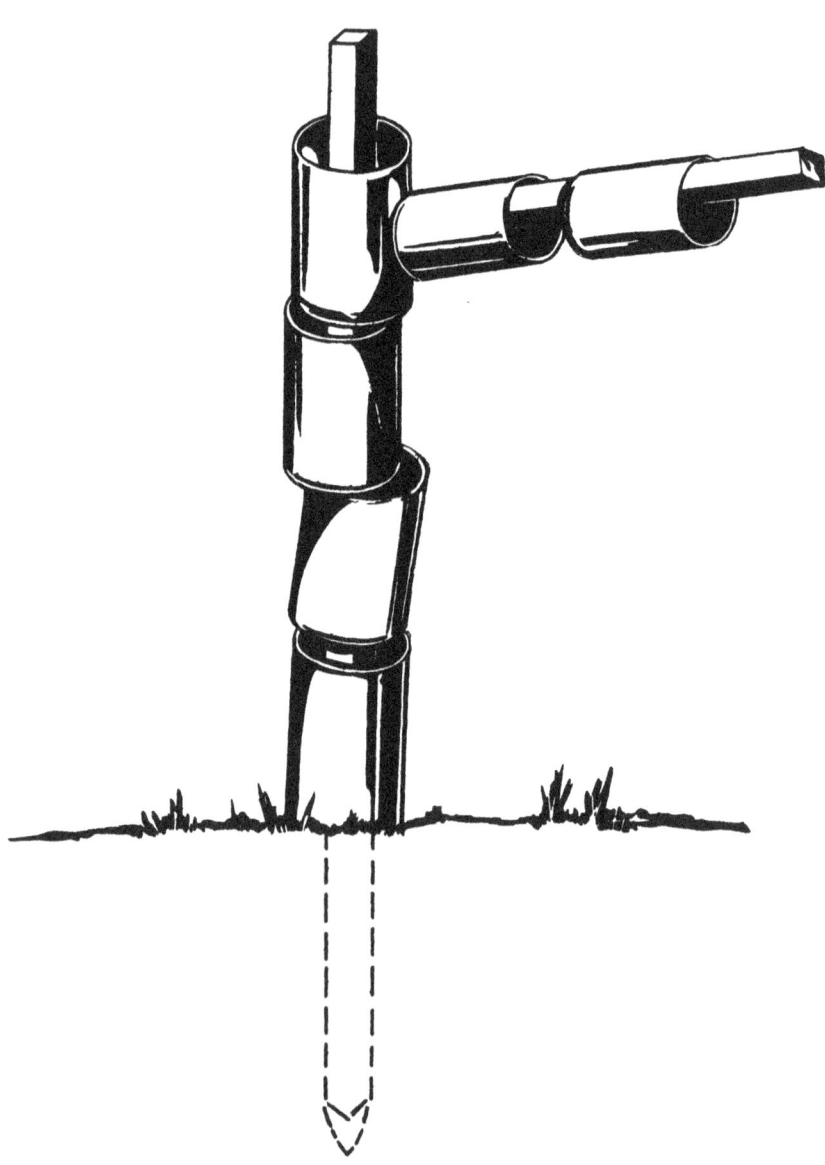

Figure 27. Use of cans or kegs supported by center post to make a realistic trunk.

Figure 28. Mobile gravel filled box for false trees.

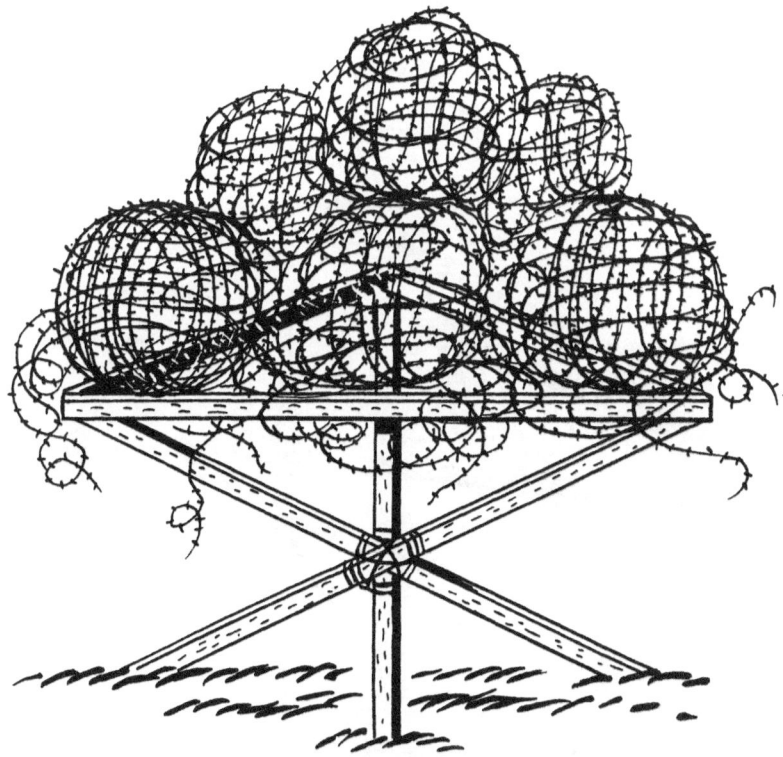

Figure 29. Tripod skeleton and barbed wire form useful in simulating a tree.

Figure 30. Low, movable stand useful in simulating a tree.

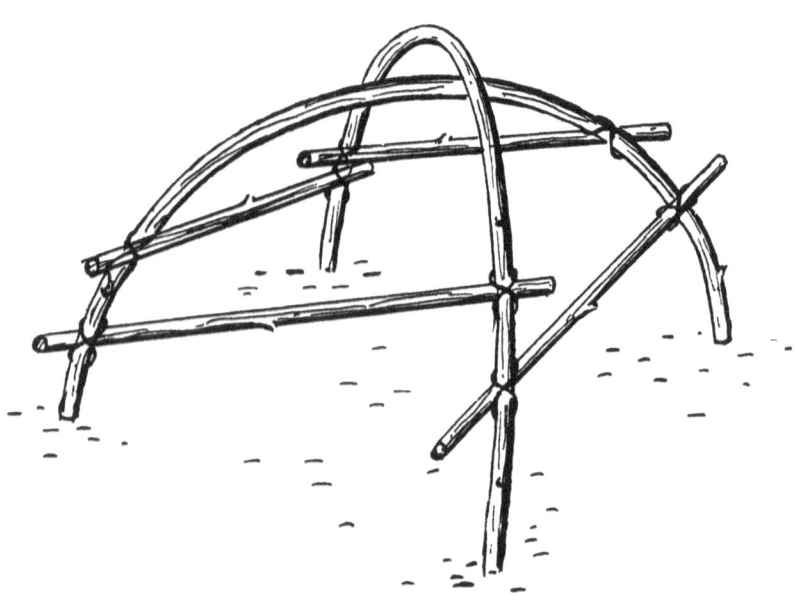

Figure 31. Sapling skeleton for simulated bush.

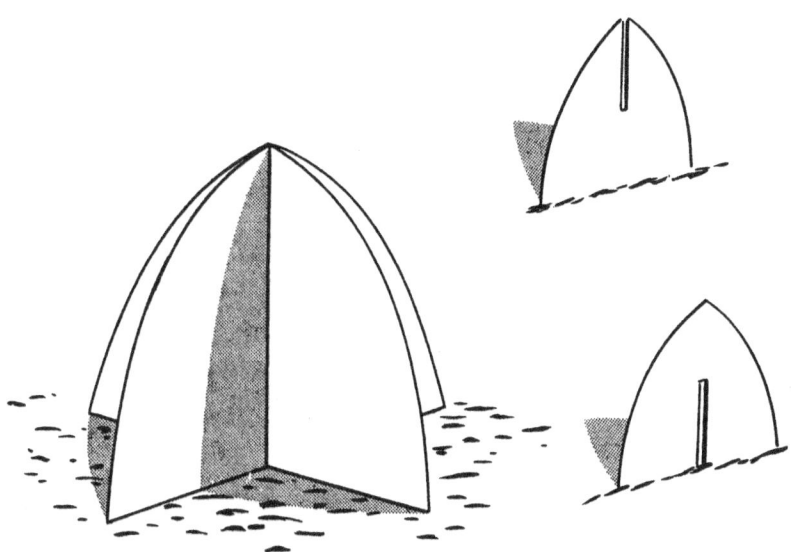

Figure 32. Cardboard or plywood silhouettes.

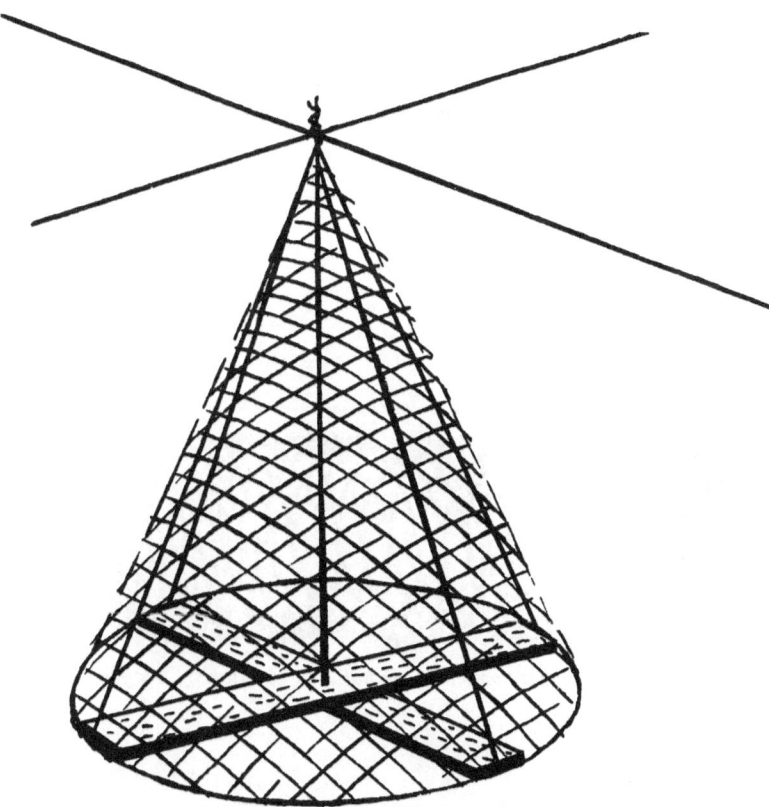

Figure 33. Suspended large cones made of wire netting augment natural overhead concealment.

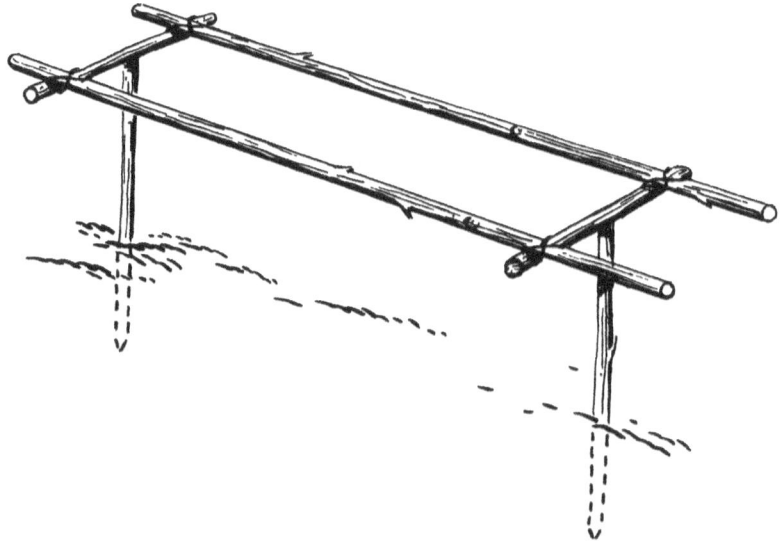

Figure 34. False-hedge framework constructed of saplings wired together.

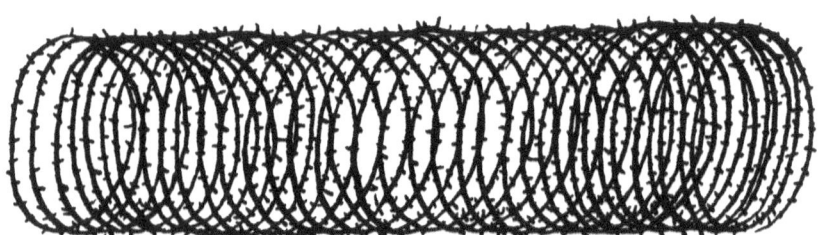

Figure 35. Wire concertina useful in simulating hedges.

priately painted and properly arranged rows of sandbags or empty gasoline cans can be used to simulate a variety of objects. Shallow ditches with the sides painted black to represent the dense shadows of deep depressions simulate drainage ditches or other excavations.

f. Grading. Grading, which is often necessary to provide adequate bomb and blast protection for small structures and fortifications, should be accomplished without interfering with concealment of the installation. Regular slopes and geometric shapes must be avoided. The final contours of the ground surface should be characteristic of the vicinity. For additional information on 3-dimensional patterning see FM 5–22.

30. Screening

Garnished netting, snow fencing, and smoke are all used as screening.

a. Large-Mesh Garnished Netting. Garnished nets can be supported in several ways, the choice depending mainly on the area to be covered, its use, and the materials available. For long spans, garnished netting can be supported by wood- or steel-truss or cable-suspension systems (fig. 36). For shorter spans, simple post-and-wire construction is adequate (fig. 37). In designing a net-support system, consideration must be given to anticipated

Figure 36. Garnished netting supported over long spans.

Figure 37. Supporting netting over short spans

snow, sleet, and wind loads. Camouflage netting must be treated as a solid surface in computing loads.

b. Snow Fencing. Snow fencing of the wood-lath type, sometimes called sand fencing, is used like garnished netting. However, the supporting structure must be stronger because of the greater weight of the fencing. Because snow fencing does not deteriorate on exposure, it is especially suitable for seacoast camouflage construction.

c. Smoke Screens. Smoke has been used extensively and successfully to conceal important installations such as docks, bridges, harbors (figs. 38 through 40), and railheads from enemy aerial reconnaissance and attack. Wind and terrain affect the laying of smoke screens. For good coverage from smoke generators, the wind should be between 2 and 14 miles per hour and the terrain should be open and fairly flat. To prevent photographic observation, a smoke screen must be highly concentrated, although it need not cover much more than the target itself. For protection against attack, however, the screen must cover an area much larger than the target, although the density can be somewhat less than that required for concealment from photo reconnaissance. For additional information on smoke screening see FM 3-5, FM 5-20, and FM 5-22.

Figure 38. Laying a smoke screen with mechanical smoke generator M1.

Figure 39. Beginning a smoke screen over a large harbor.

Figure 40. The harbor (fig. 39) completely obscured fifteen minutes later.

CHAPTER 5
REAR-AREA INSTALLATIONS

31. Camouflage of Buildings

The three basic methods of concealment—blending, hiding, and deceiving—can be applied either to existing buildings or to new construction. However, concealment is much more easily obtained when the camouflage scheme is incorporated in the designs for new projects and when structures are sited and dispersed to take advantage of terrain features and natural covers or blended or disguised to fit into an urban pattern. Camouflage of existing buildings, on the other hand, is a difficult problem if the need for concealment was not considered initially.

32. Individual Buildings and Tents

a. Disruptive or Pattern Painting. The shape and, to a limited degree, the shadow of an existing building can be disrupted by pattern painting the walls, the roof, and the ground around the structure. To accomplish this, large irregular patterns of 2 or 3 colors which simulate the local pattern are applied in such a way that the straight edges of the building are broken up. The pattern applied to the roof is darker than that on the walls, because the roof reflects more light. For this reason, if possible, roofs should be textured before being painted. The dark patterns on the roof are carried down onto the wall surface to break the line of the structure. The ground can be sprayed with black bituminous emulsion or the methods in *c* and *d* below may be used.

b. Digging In. If the terrain permits, a new structure may be partially dug in. Good blending is thus achieved, because the height and the resulting shadows are lessened. Figure 41 illustrates a hut blended into terrain by a combination of digging in, texturing, and the use of netting. Buildings constructed of metal or housing metal materials may be concealed from radar and passive infrared detectors by piling dirt in a gradual slope up to the eaves and placing approximately 3 inches of soil on the roof. Care in maintaining adequate moisture content must be exercised. This is true especially of the thin layer on the roof. The moisture content of the soil must be kept at approximately the same level as that of the surrounding soils, or a reflectance differential will be created to degrade the concealment value.

c. Ground Mats of Coarse-Textured Materials. Course-textured materials such as cinders, slag, or coal-washer refuse may be spread matlike around a building in an irregular pattern to obscure the shadow.

d. Transplanting. Transplanting thick shrubbery or trees in the shadow at the sides of a building is an effective method of absorbing the shadows of buildings.

e. Silhouettes. Silhouettes of plywood or other materials, attached to eaves, break the identifying shadows (fig. 42).

f. Screening. Buildings can be concealed by screens of garnished nettings. Figure 43 illustrates extension of the jungle through the use of garnished netting and completely conceals an operations building. In figure 44 netting garnished with steel wool distorts the shape of the building from aerial observation. On small buildings where the slope of the roof is 30° or less, the netting runs completely round the building, starting at the eaves and extending only far enough beyond the eaves to mask the ground line of the building when observed from an angle 30° above ground level (fig. 45). Where concealment from closer observation is required the netting should be gradually sloped to ground level. Disruptive patterns may be painted over netting, roof, and gable-end walls. For structures with roofs steeper than 30° the netting must cover the building.

Figure 41. Quonset hut blended into terrain.

Figure 42. Rigid silhouettes added to eaves of barracks buildings to distort building shape.

a. *Disguise.*

   (1) *Large buildings.* The nature and size of an existing large building can be disguised by making it look like several small houses (fig. 46). Shadows or trees between the small houses are simulated by texturing or painting. The parts of the roof simulating small houses are painted in colors like those on the roofs of nearby houses. All walks or paths leading directly to simulated entrances of the building should be textured so that they blend with the surrounding terrain. Depot warehouses (1, fig. 47), power houses, and similar existing large structures in urban areas are often made to appear to be a group of smaller buildings characteristic of the vicinity (2, fig. 47). This can be accomplished by the following:

   (a) Roof lines can be varied through the use of wooden framework covered with burlap on fine-mesh wire netting to simulate sloping or hip roofs.

   (b) Where paving prevents transplanting real shrubbery near a structure, 2-dimensional painting or 3-dimensional false shrubbery of steel or glass wool may be used to disrupt the shadow.

Figure 43. Concealed operations building

Figure 44. Shape of building distorted through use of garnished netting.

 (*c*) Fire walls projecting above a roof can be transformed into the garden walls or hedgerows on simulated property lines.

 (*d*) Monitors, elevator shafts, and other projections above a roof can be disguised as small buildings.

 (*e*) Garnished wire netting on saw-tooth construction can be used to conceal shadows and reflections from glass skylights.

(2) *Hutments.* Hutments can be made to conform to the architectural type common to the vicinity by erecting superstructures of light timber and covering them with burlap, plaster, or other materials. Figure 48 illustrates a quonset hut being altered to resemble a native house. The superstructure framework is covered with chicken wire which in turn will be covered with a light coat of plaster.

(3) *New structures.* New construction such as barracks can be designed in the style of local buildings (fig. 49).

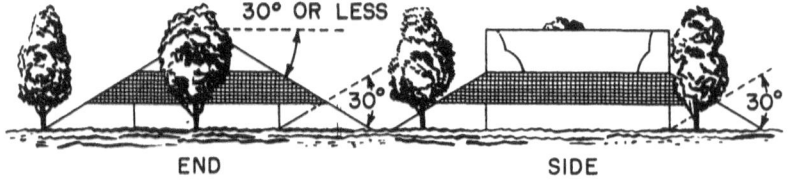

Figure 45. Diagram showing netting attached to eaves of small buildings.

Figure 46. Long military structures given the appearance of smaller local buildings

1 Warehouse-type building prior to camouflage.
2 Warehouse camouflaged.

Figure 47. Camouflage of warehouse-type building

Figure 48. Quonset hut being altered to resemble native houses in vicinity

Designs of this nature afford excellent camouflage and do not require camouflage maintenance; camouflage discipline, however, must be strictly observed to avoid giving away the military nature of the installation.

(4) *Tents.* Several methods can be used to disguise or conceal tents, with the end objective to break up the straight line effect of the tents and to disrupt the shadows they cast. One method is to erect a frame over the tent and use feather-garnished wire to simulate a tree as illustrated in figure 50. Another method is to place false trees in their growing state at various intervals around the tent as illustrated in figure 51. If more elaborate construction is necessary a frame constructed of 1 x 2 inch lumber and suspended from the crown of the tent by wire can be erected and false trees placed as illustrated in figure 52. This method requires considerable construction, but very little maintenance.

33. Grouped Buildings

a. Existing Buildings. The camouflage used for existing grouped buildings may consist of tonedown painting of the buildings, in which case it may also be necessary to tone down surrounding areas to eliminate strong color contrasts. A higher degree of camouflage may include disruptive painting (fig. 53). 1, figure 53 illustrates a large factory and grounds painted in disruptive patterns to conceal the nature of buildings from aerial observation; 2, figure 53 illustrates an aerial view of the same factory (in circle) which shows how the disrupted pattern painting has disrupted the regular shape of the factory. In figure 54 disruptive pattern painting is used in conjunction with false structures. Houses similar to those in the background of the figure are simulated by 3-dimensional false roofs to aid in camouflaging the area. However for complete concealment the installation and adjacent areas may be covered with netting supplemented with 2- and 3-dimensional patterns added to it. 1, figure 55 illustrates a catwalk provided for inspection and maintenance of camouflage and 2, 55, points out the doorway entrance to the building. The large factory building area is now completely obscured by the addition of camouflage and looks like a residential area. Figure 56 is another example of elaborate camouflage partially concealing factory buildings and areas.

b. New Construction. Less difficulty is experienced in providing adequate concealment for new construction designed and built

Figure 49. Barracks building modified to conform to style of local construction

Figure 50. Tent covered with feather-garnished wire to simulate a tree

Figure 51. Tent covered with false tree of painted tumbleweed.

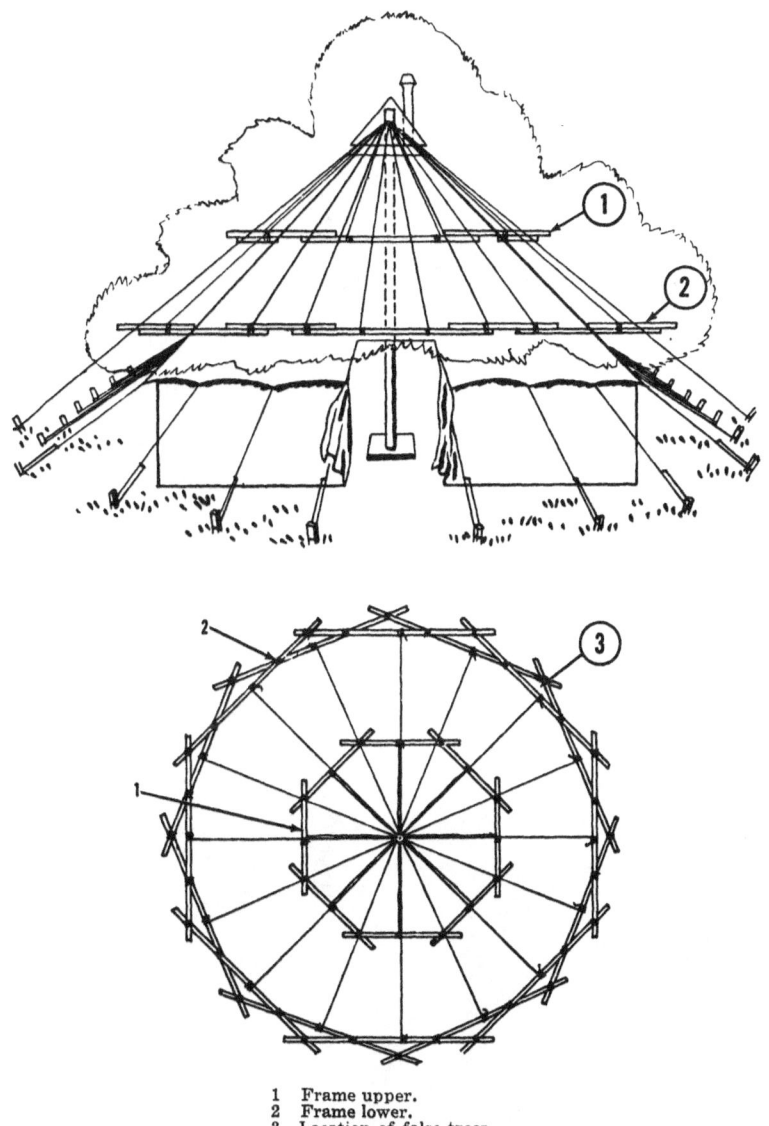

1 Frame upper.
2 Frame lower.
3 Location of false trees.

Figure 52. Supporting frame for false tree.

1 Large factory and ground painted in disruptive patterns.
2 Aerial view of same factory (in circle) showing complete camouflage.

Figure 53. Disruptive patterns painted to disguise factory

1 Inspection and maintenance catwalk.   2 Concealed building entran

Figure 54. Houses simulated by simple 3-dimensional false roofs.

Figure 55. Large factory buildings and area camouflaged to simulate residential area

Figure 56. Partial concealment of factory buildings and areas

with camouflage requirements in mind. The following basic methods have proved effective:

- (1) Dispersion in wooded areas (1, 2 and 3, fig. 57) combined with tonedown, texturing of roofs, or pattern painting.
- (2) Disguise, by following the scale, style, and layout of local architecture (fig. 49). Painting, planting, and other details must be carefully copied from the original. Concealment must be provided for military vehicles and equipment used by the occupying forces. Native workmen using local materials can be employed to insure correctness of detail.

34. Land Communications

a. Roads and Parking Areas. Roads can be effectively concealed or screened-over for only comparatively short stretches (fig. 3). However, when intersections, traffic circles, and short access roads constitute landmarks, the camouflage plan must include some scheme to make them inconspicuous (fig. 58). The following six methods are commonly used.

- (1) Tonedown painting to reduce the distance from which a road can be seen. The effectiveness of this method depends on how closely the texture and color of the treated road surface match those of adjacent areas. In painting roads, the color should be extended irregularly beyond the road edges and across the road itself.
- (2) Texturing the surface to get a truer tone value than by painting alone.
- (3) Screening with garnished netting. This provides the best concealment but is normally practical only for spur roads serving vital installations and parking areas (1 and 2, fig. 59).
- (4) Transplanting trees. This is only practical on roads serving permanent installations. Trees should be situated to conform to the pattern of the area and to cast the maximum amount of shadow on the road. Remaining gaps should be concealed by overhead screens.
- (5) Relocating roads. Straight lines may be replaced by curves which follow contours and make maximum use of natural cover; two-lane roads may be redesigned as separated single-lane roads.
- (6) Making decoy roads by scarring (fig. 60), shallow ditch-

1 Typical military barracks.
2 Dispersed barracks.
3 Barracks constructed with camouflage considered.

Figure 57. Camouflage of barracks.

Figure 58. Short access road through cut effectively screened.

ing, texturing, or spreading out lengths of fabric (fig. 61). This tends to direct enemy attention away from camouflaged areas.

b. *Railways, Railheads, and Rolling Stock.*
  (1) *Railways.* Railways are recognized from the air by long parallel lines, gradual curves, light-colored stone ballast in roadbeds, and by shine from mainline tracks which is sometimes glaring to oblique aerial observation. Ballast can usually be toned down by spraying it with paint to blend it with the surrounding vegetation or ground pattern. The bright surface of rails can be dulled by applying diluted solutions of most acids, especially hydrochloric and sulfuric acids. False railway lines can be used in operational camouflage schemes and in decoys for ear-area installations and railway artillery. False rails may be constructed of various materials such as 2- x 4-inch lumber (fig. 62). Ties may be simulated by ground painting (fig. 62) or by lengths of scrap tin (fig. 63), lumber, or cardboard.
  (2) *Railheads.* Railheads are difficult to conceal because of the activity connected with them and the large areas they cover. Camouflage measures may be used under ideal conditions to conceal the installations and under adverse conditions to make the target appear unprofitable to the enemy. Successful concealment depends largely on the site. The site should provide as much natural overhead cover as possible for spur tracks, truck

1 Exterior view
2 Interior view.

Figure 59. Parking area screened from observation.

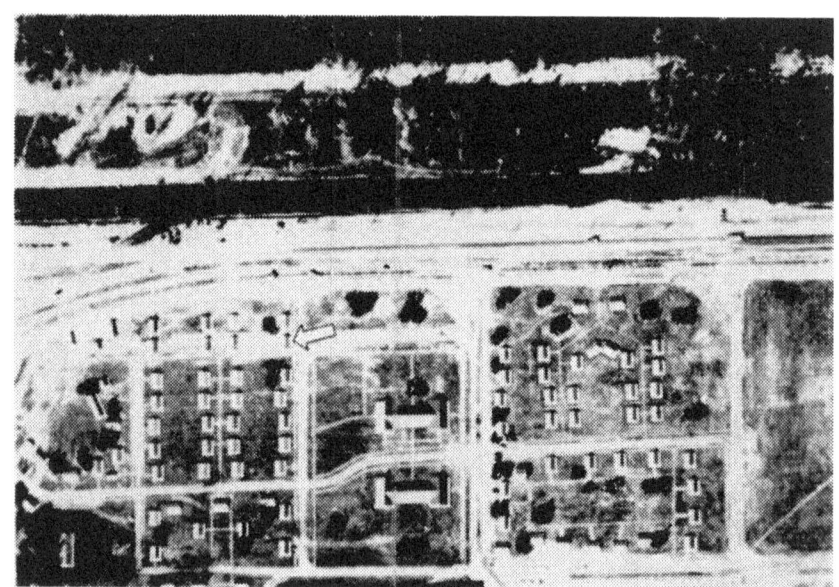

Figure 60. A decoy housing area with decoy roads

Figure 61. False road over top of dug-in position.

Figure 62. Rails simulated by 2- x 4-inch lumber coated with aluminum paint.

parks, storage areas, warehouses, and access routes. The railhead itself should be at some intermediate point, not at the end of the rail line. The junction with the mainline should be made as inconspicuous as possible. Specific camouflage techniques in addition to those mentioned in (1) above are:

(a) Adding extra ballast to cover ties.
(b) Making outer edges of ballast irregular.

Figure 63. Ties and rails made from flattened and shaped tin.

Figure 64. Fitted garnished netting used to conceal ties and ballast

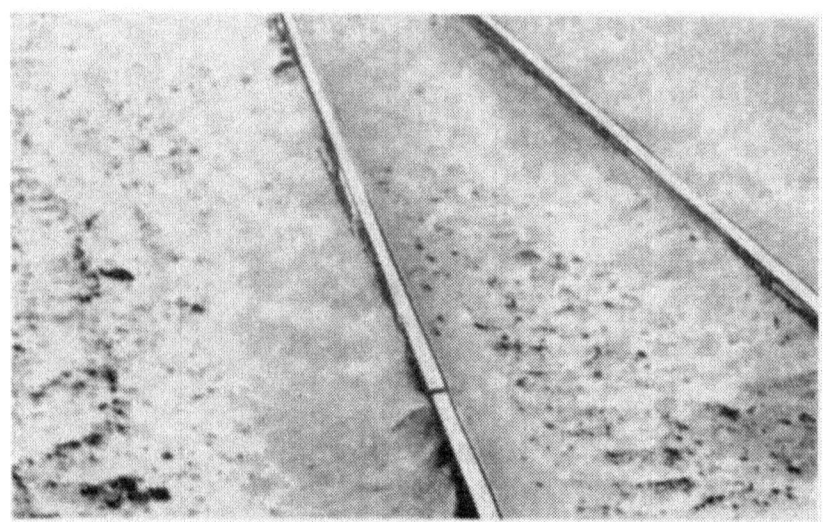

Figure 65. Decoy railway line in desert area.

(c) Placing fitted screening between and on sides of rails (fig. 64).

(d) Erecting nets over sidings, between buildings to conceal loading platforms, or over access routes to storage areas. Where steam locomotives are used, the netting must be high enough to avoid the danger of fire caused by the escape of sparks or hot cinders from the stack. Coal-burning locomotives must also be equipped with spark arresters on smoke stacks and should have double ashpans. Diesel, gasoline, and other self-propelled rail vehicles should have spark arresters properly installed on exhaust stacks.

(e) Enforcing a track plan for vehicles. This will keep visible signs of unrelated activities at a minimum.

(f) Providing camouflage for supplies stored in the open.

(g) Building and simulating the operation of a decoy railhead. Whether or not the railhead can be concealed, this method, skillfully carried out, is probably the best solution for defeating or confusing enemy observation and attack. A section of false track in a decoy railway leading to a decoy railhead is shown in figure 65. A decoy train erected at the same railhead is shown in figure 66.

(h) Dispersing freight cars, locomotives, vehicles, and supplies. This is a normal precaution and is essential when other camouflage is impractical.

(3) *Rolling stock.* Locomotives and tank cars offer profitable targets to low-flying planes. Figures 67 and 68 show how a Diesel locomotive can be made to look like a boxcar by adding a superstructure of mild-steel bar uprights and sheet panels. Tank cars may be similarly disguised (figs. 69 and 70).

c. Bridges. Bridges are extremely difficult to camouflage. However, a number of camouflage devices can be used to deceive the enemy as to the location and condition of a bridge.

(1) The enemy may be deceived by simulated damage (fig. 71). Simulated craters may be painted on the decking, covered with a suitable material (2, fig. 71), and exposed after an attack on the bridge (2, fig. 72). Portable false tanks (4, fig. 72) or other vehicles may be placed to simulate a traffic jam after the attack. Portions of the

Figure 66. Decoy railway train on same railway shown in figure 65

Figure 67. Framework used in disguising Diesel locomotive to resemble boxcar

75

Figure 68. Diesel locomotive disguised as a boxcar.

side railing may be removed (2, fig. 72). Another method of making the bridge appear damaged is to prepare an overhead or vertical framework of wire or cable (1 and 2, fig. 71) on which debris is stocked in readiness and spread in an irregular manner after an attack (1 and 2, fig 72). For this type of deception to be convincing, it must appear that an alternate river crossing has been or is being prepared. Approaches to this decoy must be well worn on both banks. If there is a line of craters on the shore, one or more must be filled. Figure 73 is an aerial photograph of a decoy bridge.

(2) A destroyed bridge may be made to appear repaired and restored for use by filling a gap with a wire or cable framework covered with cloth or other material. Thus, it again becomes a logical target and protects the actual crossing.

(3) Large bridges may be concealed by smoke (par. 30).

(4) Where the riverbed is suitable and the water slow and muddy, bridges may be constructed with the deck submerged just below the surface.

(5) The shape and shadow of a destroyed bridge may be used to help conceal a ponton bridge constructed alongside. If a decoy crossing is built at a logical place some

Figure 69. Framework for boxcar used to disguise European tank car.

Figure 70. Tank car disguised as boxcar.

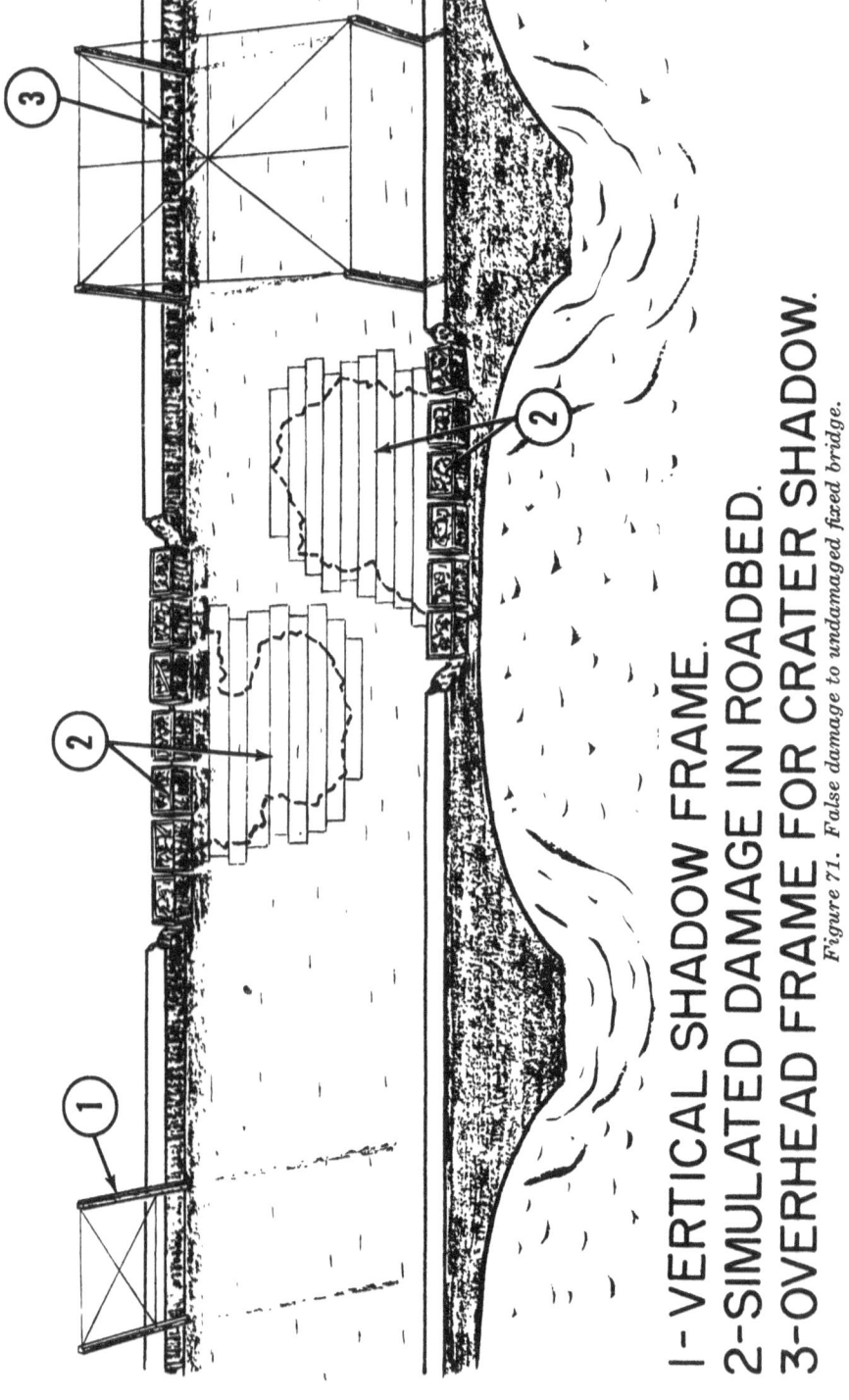

Figure 71. False damage to undamaged fixed bridge.

1- VERTICAL SHADOW FRAME.
2- SIMULATED DAMAGE IN ROADBED.
3- OVERHEAD FRAME FOR CRATER SHADOW.

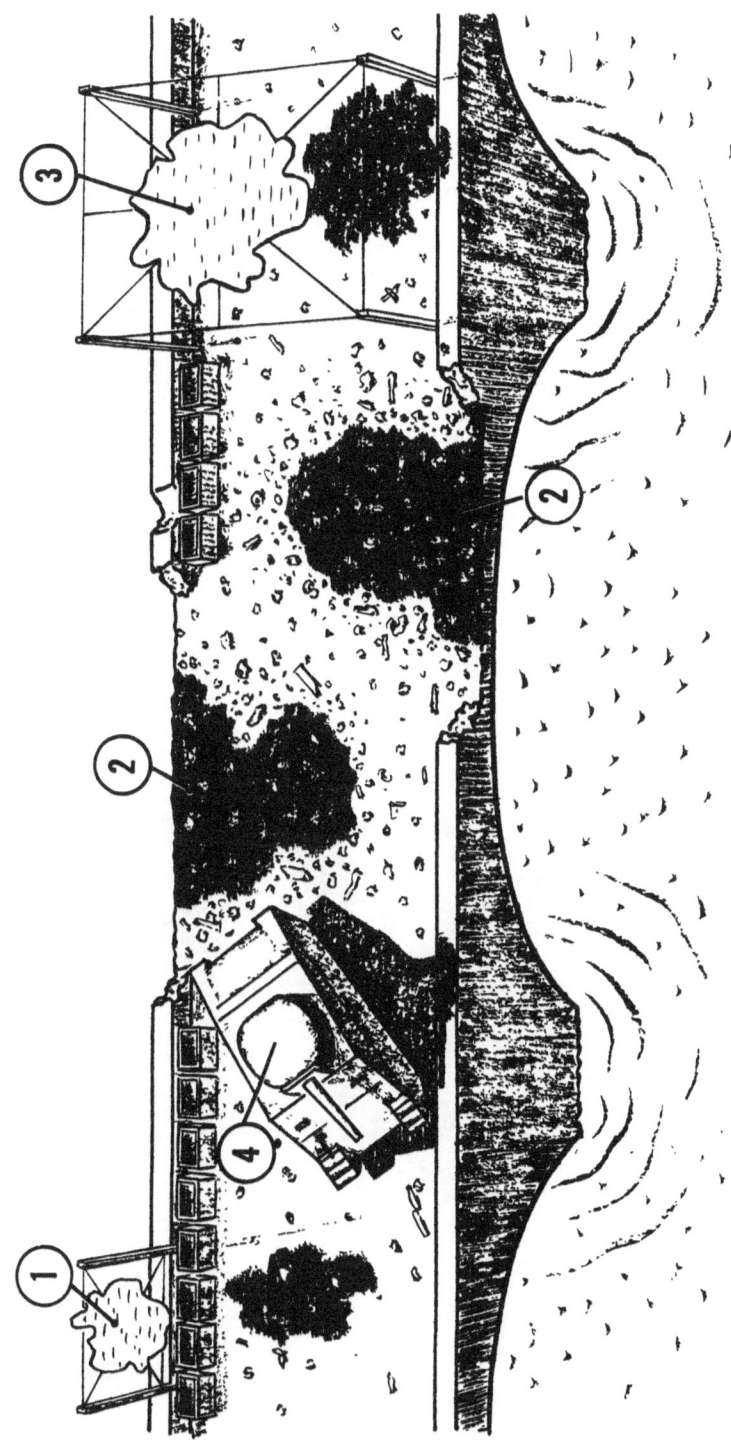

1 Vertical shadow frame; debris attached to simulate crater shadow.
2 Simulated damage exposed and railing removed.
3 Debris attached to overhead shadow frame to simulate crater shadow.
4 False tank.

Figure 72. False damage exposed after attack.

Figure 73. Aerial view of decoy ponton bridge

distance away, the expedient has a greater chance of escaping detection.

(6) Occasionally construction and subsequent use of a bridge may be concealed from enemy ground observation by vertical screens suspended from poles and cable supports.

d. *Wire and Pipe Lines*.

(1) *Wire lines*. A well-camouflaged structure is of little value if a conspicuous line of communication wire terminates at the installation. For example, it is disastrous to allow unconcealed cable lines to end abruptly at what is meant to appear as an innocent hill but is actually an important command post. A decoy must continue past the installation to a logical termination or the real line must be camouflaged if above ground. Imitation lines can be made of rope, wire, cord, or other similar materials. The presence of a line can be concealed to a great extent by carefully locating it along terrain lines. Irregularly sized supporting poles with the bark left on, set at irregular intervals and staggered to conform to the ground pattern, are less conspicuous than lines regularly spaced and alined. Spoil taken from the pole holes must be carried away or hidden. Care must be taken during maintenance to avoid making an obvious path along the line of poles.

(2) *Pipe lines*. Pipe lines should be laid along secondary roads wherever possible. When cross-country laying is necessary, terrain features should be fully utilized. To eliminate the shadow of the pipe, dirt or debris blended with the background should be banked gently along both sides of the pipe. A tonedown color applied to the pipe helps blend it with the background. Tanks and pumping equipment should be recessed in pits, dispersed, and concealed by natural cover or nets. False pipe lines are easily simulated by the use of ditching equipment; after each day's work, several poles should be left at the end of the ditch to simulate a stack of unlaid pipe.

35. Concealment of Water

a. Reason for Concealment. Waterways and lakes which may serve as reference points to the enemy in the attack on specific targets may have to be concealed to hide their shine. In connection with large-scale camouflage plans for important installations, it may even be necessary to simulate a landscape over a water area.

The construction method is determined by the reason for hiding the water area, its size, the character of the bottom, and the characteristics of the water area. Water areas are characterized as being still, swift-flowing, tidal, deep, or shallow.

  b. Methods of Concealment.

   (1) *Use of floats.* Floats can be constructed using the following materials:

   (a) In tropical fresh water, rafts of green bamboo will sprout and cover the area with natural green foliage.

   (b) Woven reed matting, treated with a preservative such as tar or heavy road mix and held in place by a log boom, will float for a long time.

   (c) Floats of watertight drums, anchored and connected by boards or wire, will support a covering material.

   (2) *Use of netting.* Netting can be supported on wire or cable strung from side to side of narrow waterways (fig. 74). If the distance is too great for a single span, sections of the framework may be supported on piles, moored rafts, or small boats.

   (3) *Draining.* Draining is possible only in the case of ponds or unused canals and often requires constant pumping.

36. Deliberate Minefields

The basic principles of camouflage outlined in FM 5-20 must be followed in concealing a minefield. The problem of concealing a minefield is not only the concealing of the mines themselves but also the boundaries and the pattern of the minefield as well. Mines must be dug in, spoil removed, and sod, snow, or other material replaced over the mine insuring that it blends in with the environment. Particular care must be taken to eliminate any hump or depression after burying the mines because this will not only disclose the minefield but the pattern of the minefield as well. Vehicle and personnel tracks must be brushed out or concealed in some other manner. The ground must appear undisturbed. The use of deception in connection with minefields is almost limitless. A decoy minefield can be as effective an obstacle as a real minefield because the enemy must dig up each simulated mine to insure it is not a real mine. For additional information on decoy minefields, see FM 5-23.

SCREEN OVER COVE

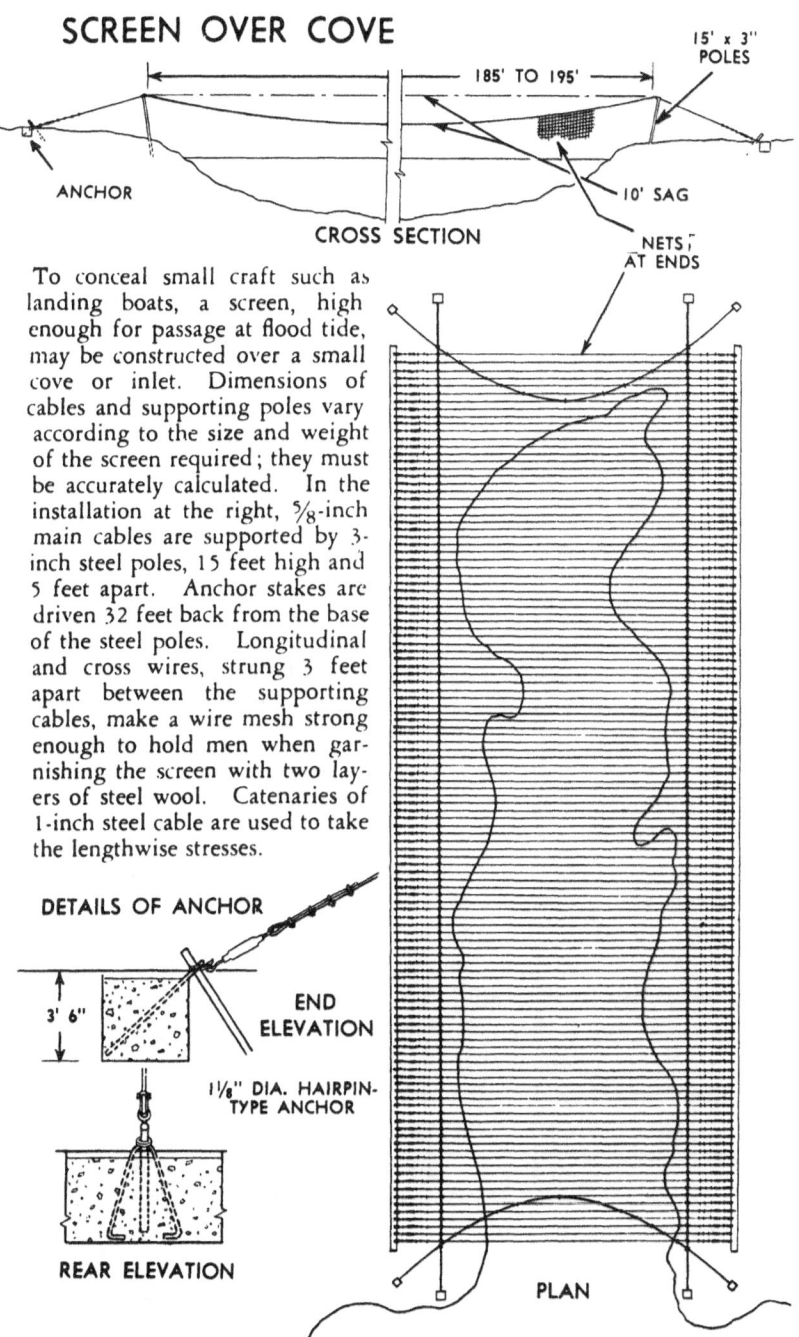

To conceal small craft such as landing boats, a screen, high enough for passage at flood tide, may be constructed over a small cove or inlet. Dimensions of cables and supporting poles vary according to the size and weight of the screen required; they must be accurately calculated. In the installation at the right, $\frac{5}{8}$-inch main cables are supported by 3-inch steel poles, 15 feet high and 5 feet apart. Anchor stakes are driven 32 feet back from the base of the steel poles. Longitudinal and cross wires, strung 3 feet apart between the supporting cables, make a wire mesh strong enough to hold men when garnishing the screen with two layers of steel wool. Catenaries of 1-inch steel cable are used to take the lengthwise stresses.

Figure 74. Screen over cove.

CHAPTER 6
MAINTENANCE

37. General

If camouflage maintenance in large installations is continuous or frequent enough to prevent deterioration, a minimum of labor and materials is required. The labor and materials required also depend greatly on the care used during the design stage and in selecting materials and on methods of application.

38. Specific Recommendations

a. Aerial Checking. A regular schedule of aerial checking should be followed. Observation flights and aerial photographs must be taken from the same elevations and directions as previous flights and photographs, so that changes in appearance can be detected readily and corrected immediately.

b. Paint. The type paint, the surface to which it is applied, the weather, and traffic affect the time in which the color fades or changes noticeably. An expedient paint-spraying device useful on large areas where different colors are involved (fig. 75) makes possible transporting a number of different colors, any one of which can be used at will. Hose lengths may be up to 200 feet, depending on the size of the compressor used.

c. Nets and Cables. Nets and screens require periodic structural maintenance. Abnormally high winds or snow loads weaken supporting systems which should be checked frequently during and after such strain. During times of unusual strain, additional temporary supports may prevent collapse or serious damage. When cables are used to supporting netting, all cable clamps must be examined and tightened about every 2 weeks until the cable has settled into shape. Otherwise, the clamps become loose and the cables slip.

d. Trees and Shrubs. Trees and shrubs require considerable care after transplanting. Supplementary concealment measures may be necessary when leaves fall from deciduous trees and their cover becomes inadequate.

e. Signs of Activity. False roads, paths, houses, and other evidence of occupation must appear to be used. Roads and paths which do not show signs of wear are not convincing. Houses that develop sagging roofs are equally unconvincing. The original traffic plan should be followed. If new tracks must be made they must be covered or made to look like old tracks.

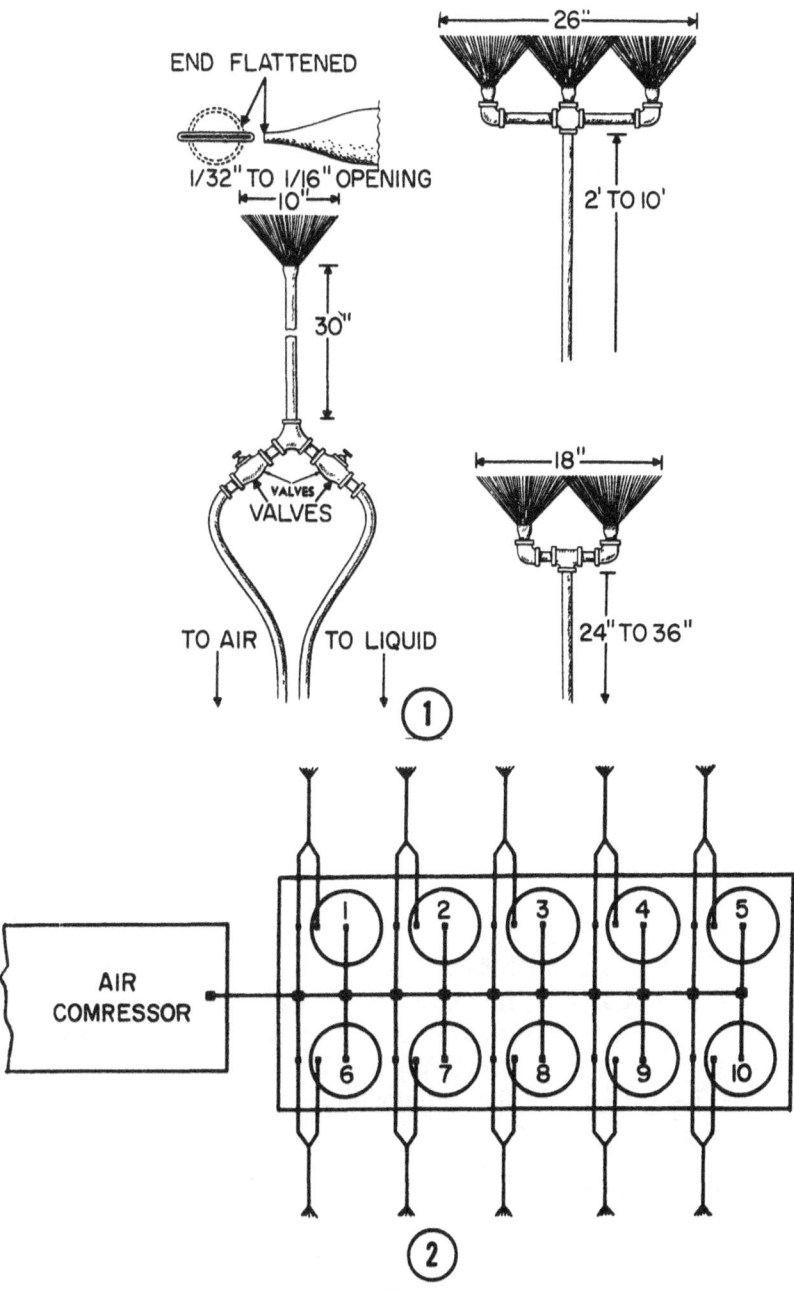

1 Nozzle arrangement.
2 Diagram of arrangement of paint set up.

Figure 75. Expedients for large scale maintenance painting.

APPENDIX
REFERENCES

1. Related Technical Publications

(0)	FM 3-5	Tactics and Techniques of Chemical, Biological, and Radiological Warfare.
	FM 5-15	Field Fortifications.
	FM 5-20	Camouflage, Basic Principles and Field Camouflage.
	FM 5-22	Camouflage Materials.
	FM 5-23	Field Decoy Installations.
	FM 5-34	Engineer Field Data.
	FM 21-5	Military Training.
	FM 21-6	Techniques of Military Instructions.
	FM 21-26	Map Reading.
	FM 21-30	Military Symbols.
(0)	TM 30-245	Photo Interpretation Handbook.
	TM 30-246	Tactical Interpretation of Air Photos.

2. Military Terms, Abbreviations, and Symbols

AR 320-5	Dictionary of United States Army Terms.
AR 320-50	Authorized Abbreviations

3. Publication Indexes

DA Pam 108-1	Index of Army Motion Pictures, Film Strips, Slides and Phono-recordings.
DA Pam 310-1	Index of Administrative Publications.
DA Pam 310-3	Index of Army Training Publications.
DA Pam 310-4	Index of Technical Manuals, Technical Bulletins, Supply Bulletins, Lubrication Orders, and Modification Work Orders.
GTA 5-1	Camouflage and Concealment.

[AG 384.6 (30 Oct 58)]

By Order of *Wilber M. Brucker*, Secretary of the Army:

MAXWELL D. TAYLOR,
*General, United States Army,
Chief of Staff.*

Official:
    R. V. LEE,
*Major General, United States Army,
    The Adjutant General.*

Distribution:
Active Army:

CNGB (2)	5-59 (1)
Technical Stf, DA (2) except	5-67 (1)
CofEngrs (25)	5-96 (10)
USA Arty Bd (2)	5-97 (5)
USA Armor Bd (2)	5-137 (1)
USA Inf Bd (2)	5-138 (1)
USA Air Def Bd (2)	5-139 (1)
USA Abn & Elct Bd (2)	5-157 (1)
USA Avn Bd (2)	5-167 (1)
USCONARC (5)	5-192 (3)
US ARADCOM (2)	5-216 (1)
US ARADCOM Rgn (2)	5-217 (1)
OS Maj Comd (10)	5-218 (1)
OS Base Comd (5)	5-226 (1)
Log Comd (2)	5-227 (1)
MDW (2)	5-262 (1)
Armies (10)	5-266 (1)
Corps (5)	5-267 (1)
Div (10)	5-278 (1)
Engr Gp (1)	5-279 (1)
Engr Bn (1)	5-301 (1)
USMA (10)	5-312 (2)
Svc Colleges (2)	5-316 (2)
Br Svc Sch (2) except	5-317 (1)
USAES (462)	5-324 (1)
Trans Terminal Comd (2)	5-328 (1)
Engr RD Lab (2)	5-329 (1)
Engr Sup Con Ofc (5)	5-344 (1)
Div Engr (1)	5-346 (2)
Engr Dist (1)	5-347 (1)
Mil Dist (2)	5-348 (1)
USA Corps (Res) (2)	5-349 (1)
Sector Comd, USA Corps	5-359 (2)
(Res) (2)	5-367 (2)
Units org under fol TOE:	5-387 (1)
5-16 (1)	5-412 (4)
5-17 (1)	5-416 (1)
5-36 (1)	5-417 (2)
5-37 (1)	5-500 (AA-AC) (1)
5-48 (1)	5-526 (1)
5-56 (1)	5-527 (2)
5-57 (1)	

NG: State AG (3); units—same as Active Army except allowance is one copy to each unit.

USAR: Same as Active Army except allowance is one copy to each unit.
For explanation of abbreviations used, see AR 320-50.

www.ingramcontent.com/pod-product-compliance
Lightning Source LLC
Chambersburg PA
CBHW022109160426
43198CB00008B/407